S0-CAS-589

pK_a Prediction for Organic Acids and Bases

pK_a Prediction for Organic Acids and Bases

D. D. Perrin

John Curtin School of Medical Research
Australian National University
Canberra

Boyd Dempsey and E. P. Serjeant

Faculty of Military Studies
University of New South Wales
Royal Military College
Duntroon

London and New York

CHAPMAN AND HALL

First published 1981 by
Chapman and Hall Ltd
11 New Fetter Lane, London EC4P 4EE
Published in the USA by
Chapman and Hall
in association with Methuen, Inc.
733 Third Avenue, New York NY 10017

Printed in Great Britain at the
University Printing House, Cambridge

ISBN 0 412 22190 X

British Library Cataloguing in Publication Data

Perrin, D. D.
pK_a prediction for organic acids and bases.
1. Acids, Organic 2. Bases (Chemistry)
3. Ionization
I. Dempsey, Boyd II. Serjeant, E. P.
547.1′3723 QD477

ISBN 0-412-22190-X

Contents

Preface

Many chemists and biochemists require to know the ionization constants of organic acids and bases. This is evident from the Science Citation Index which lists *The Determination of Ionization Constants* by A. Albert and E. P. Serjeant (1971) as one of the most widely quoted books in the chemical literature. Although, ultimately, there is no satisfactory alternative to experimental measurement, it is not always convenient or practicable to make the necessary measurements and calculations. Moreover, the massive pK_a compilations currently available provide values for only a small fraction of known or possible acids or bases. For example, the compilations listed in Section 1.3 give pK_a data for some 6 000–8 000 acids, whereas if the conservative estimate is made that there are one hundred different substituent groups available to substitute in the benzene ring of benzoic acid, approximately five million tri-substituted benzoic acids are theoretically possible.

Thus we have long felt that it is useful to consider methods by which a pK_a value might be predicted as an interim value to within several tenths of a pH unit using arguments based on linear free energy relationships, by analogy, by extrapolation, by interpolation from existing data, or in some other way. This degree of precision may be adequate for many purposes such as the recording of spectra of pure species (as anion, neutral molecule or cation), for selection of conditions favourable to solvent extraction, and for the interpretation of pH-profiles for organic reactions. Prediction is also valuable where the experimental pK_a determinations are difficult or impossible to perform, such as with weak acids and bases lacking adequate spectral differences in the acid and base forms, with substances that are unstable, or with substances that are insufficiently soluble.

The literature on pK_a prediction is scattered, but the bulk of it is comprised of two reviews dealing with organic bases (Clark and Perrin, 1964) and organic acids (Barlin and Perrin, 1966), a review of the use of pK_a values in structure elucidation (Barlin and Perrin,

1972), a paper on the prediction of the strengths of some organic bases (Perrin, 1965) and a recent lecture (Perrin, 1980). It seemed opportune to gather this material together, amplifying the methods of calculation, so as to provide a book that was a companion volume to *The Determination of Ionization Constants* and *Buffers for pH and Metal Ion Control* (Perrin and Dempsey, 1974). It is written with the same readers in mind, namely physical and organic chemists and biochemists.

The most useful methods of pK_a prediction are based on the assumption that within particular classes of acids and bases, substituents produce free energy changes which are linearly additive. This assumption has led to many published Hammett and Taft equations which are given in the Appendix together with extensive tables of substituent constants. A large number of worked examples using these equations, and other predictive methods occur throughout the text.

It is important to recognize situations where simple predictions may not be possible, perhaps because of steric or resonance interactions; some typical examples are discussed in Chapter 9. Many compounds, especially of biological origin, have complicated structures and may contain more than one acidic or basic centre. Such problems are considered, especially in Chapter 7. Throughout the book, practical rather than theoretical aspects of pK_a prediction are emphasized, but references to the literature are included which give further discussion of the theory on which the methods are based.

D. D. Perrin
Boyd Dempsey
E. P. Serjeant

Canberra
December, 1980

Chapter One

Introduction

Recent years have seen considerable effort put into quantitative explanation and prediction of the effect of substituents on physical properties such as ionization constants and on the reactivities of organic molecules. As a result it is now possible to predict to within several tenths of a pH unit many of the pK_a values which express the strengths of organic acids and bases. This book seeks to provide guidelines to enable such predictions to be made, illustrating the methods used and giving numerous worked examples.

1.1 The concept of pK_a

An acid is 'a species having a tendency to lose a proton' (Brönsted, 1923) while a base is 'a species having a tendency to add on a proton'. Hence for every acid, HA, there is a conjugate base, A^-:

$$HA \rightleftharpoons H^+ + A^- \tag{1.1}$$

and for every base, B, there is a conjugate acid, BH^+:

$$BH^+ \rightleftharpoons H^+ + B \tag{1.2}$$

Thus, acetic acid–acetate ion and ammonium ion–ammonia are examples of conjugate acid–base pairs. If HA (or BH^+) has a great tendency to lose protons, it follows that its conjugate, A^- (or B), has only a small tendency to accept protons. In other words, if HA (or BH^+) is a *strong* acid, A^- (or B) is a *weak* base, and vice versa.

Acids and bases so defined can only manifest their properties by reacting with bases and acids respectively. In aqueous solution, acids react with water acting as a base:

$$\text{HA (or BH}^+) + \text{H}_2\text{O(base)} \rightleftharpoons \text{H}_3\text{O}^+ + \text{A}^-\text{(or B)} \tag{1.3}$$

and bases react with water acting as an acid:

$$\text{A}^-\text{(or B)} + \text{H}_2\text{O(acid)} \rightleftharpoons \text{HA(or BH}^+) + \text{OH}^- \tag{1.4}$$

Quantitatively, the acid strength of HA, or BH^+, relative to the base strength of water is given by the equilibrium constant expression for Equation (1.3):

$$K = \frac{(H_3O^+)(A^- \text{ or } B)}{(H_2O)(HA \text{ or } BH^+)} = \frac{(H_3O^+)(\text{base species})}{(H_2O)(\text{acid species})} \quad (1.5)$$

where parentheses denote activities.

As almost all measurements are made in dilute aqueous solution, the concentration of water remains essentially constant and its activity can be taken as unity. Letting H^+ represent the solvated proton, we have $K_a = (H^+)(B)/(A)$, where K_a is the acidic dissociation (or ionization) constant, and B and A represent base and acid species, respectively. This equation can be written in the form

$$pK_a = pH + \log((A)/(B)) \quad (1.6)$$

where pK_a is the negative logarithm of K_a, and is equal to the pH at which the activities of A and B are equal.

Equation (1.6) also expresses the strength of the conjugate acid of an organic base (commonly spoken of as 'the pK_a of the organic base', the 'basic pK_a' or 'the pK for proton addition'). Thus $pK_a = 9.25$ for the ammonium ion lies on the same scale as $pK_a = 4.76$ for acetic acid and 10.00 for phenol. The greater the pK_a value, the weaker the substance as an acid, or, conversely, the stronger is its conjugate base. For any given solvent, the pK_a scale is convenient for expressing the strengths of both acids and bases. The earlier practice of defining the 'basic pK', pK_b, from the relation

$$K_b = (BH^+)(OH^-)/(B) \quad (1.7)$$

is thus unnecessary. Note that pK_a and pK_b are related by the equation

$$pK_a + pK_b = -\log(\text{ionic product for water, } K_w) \approx 14 \quad (1.8)$$

Table 1.1 lists typical ranges of pK_a values for some of the more common types of acids and bases, while Table 1.2 gives specific examples.

1.2 The usefulness of pK_a values

A large number of organic compounds, both natural and synthetic, contain acidic and/or basic groups which govern many of their chemical, physical and biological properties. For such compounds,

Table 1.1 *Typical ranges of* pK_a *values for organic acids, including conjugate acids of organic bases*

Classes of compounds	Typical pK_a ranges
Ethers (as bases)	−2– −4
Pyrimidinium ions	1–2
Aliphatic dicarboxylic acids	1–4.5 (1st dissociation)
Anilinium ions	1–5†
α-Amino acids	2–3 (COOH)
Monocarboxylic acids	3–5‡
$RONH_3^+$	4–5
Pyridinium ions	4–6
Aliphatic dicarboxylic acids	5–7 (2nd dissociation)
Thiophenols	5–7
α, β-Unsaturated aliphatic aminium ions	6–9
Imidazolinium ions	7
Hydroxyheteroaromatics	7–11
Phenols	8–10§
Purines	8–10
α-Amino acids	9–10.5 (NH_3^+)
Saturated nitrogen heterocycles	9–11 (NH_3^+)
Aliphatic and alicyclic aminium ions	9–11
Thiols	9–11
Oximes	10–12
Guanidinium ions	11–14
Aldehydes	11–14
Azoles	13–16
Alcohols and sugars	13–16

But note: † 2-nitroanilinium ion, pK_a = −0.3
‡ trifluoroacetic acid, pK_a = −0.26; 2,4,6-trinitrobenzoic acid, pK_a = 0.65
§ 2,4,6-trinitrophenol, pK_a = 0.22

the proportions of the species (neutral molecule, anion, cation) that are present at a particular pH are determined by the pK_a value and can be calculated from Equation (1.6), conveniently rearranged in the form:

$$\frac{(\text{base})}{(\text{acid})} = 10^{\text{pH} - \text{p}K_a} \tag{1.9}$$

In spectrophotometry it is desirable to obtain the spectrum of a particular species free from its conjugate(s). For this requirement, it follows from Equation (1.9) that the pH of the medium should be outside the range, p$K_a \pm 2$. However, if the difference between pH and pK_a is less than 2, knowing the pK_a allows calculation of the necessary corrections.

In pharmacology, the pK_a controls many aspects of drug metabolism, including transport through membranes which are

Table 1.2 *Individual* pK_a *values of organic acids and bases in water at 20–25°*

Bases	pK_a	Acids	pK_a
Pyrrole	−3.8	Methane sulphonic acid	−6.0
Indole	−2.3	Aminoacetic acid	2.35
Tetrahydrofuran	−2.1	2-Furoic acid	3.16
Urea	0.1	Formic acid	3.75
4-Pyrone	0.1	Benzoic acid	4.21
Diphenylamine	0.77	Succinic acid	4.22, 5.64
Oxazole	0.8	Acetic acid	4.76
Pyrimidine	1.23	Cyclohexanecarboxylic acid	4.90
Thiazole	2.44	Uric acid	5.83
Aniline	4.69	Thiophenol	6.52
Quinoline	4.92	*p*-Nitrophenol	7.15
Pyridine	5.23	Acetylacetone (enolic form)	8.13
Isoquinoline	5.42	Purine	8.93
Aminoacetic acid	9.78	Phenol	10.00
Triethylamine	10.78	Ethanethiol	10.54
Ethylamine	10.81	2-Pyridone	11.65
Diethylamine	11.09	Formaldehyde	13.29
Piperidine	11.28	Methanol	15.5
Acetamidine	~12.4	Pyrrole	~16.5

frequently permeable only to a particular species. In physiology, the effects of a substance of known pK_a can be considered in the light of the proportions of the various species existing under the appropriate pH conditions.

Considerations of theoretical aspects of proton addition and removal facilitate structural studies. In particular, the sites of successive proton addition and removal can be often assigned. In some cases, confirmation of structure can be made with the help of pK_a values. Likewise, the pH-profile in reaction kinetics depends closely on pK_a values.

In preparative chemistry, pK_a values can be used to select conditions for synthesis especially by considering the effects of pH on reaction products and on the properties of postulated intermediates not available for measurement. In the isolation of acids and bases from aqueous media by solvent extraction, the uncharged species is generally more soluble in the organic phase. The pH of the aqueous phase can be adjusted to its optimum value for extraction if the pK_a is known.

In analytical chemistry, pK_a values assist in the interpretation of pH titrations where multiple acidic or basic sites are present. Buffer capacities can be calculated at known pH values and buffer concentrations, provided the pK_a of the buffer acid is known. pK_a values are also required in the estimation of micro-constants and in the calculation of tautomeric equilibria.

1.3 Compilations of pK_a values

Extensive compilations of pK_a values are available. These include, in particular:

'Ionization constants (of heterocyclic substances)', Albert, A. (1963), in *Physical Methods in Heterocyclic Chemistry* (ed. Katritzky), Academic Press, New York, Vol. 1, p. 2.

Dissociation Constants of Organic Acids in Aqueous Solution, Kortum, G., Vogel, W. and Andrussow, K. (1961), Butterworths, London.

Dissociation Constants of Organic Bases in Aqueous Solution, Perrin, D. D. (1965), Butterworths, London.

Dissociation Constants of Organic Bases in Aqueous Solution, Supplement 1972, Perrin, D. D. (1972), Butterworths, London.

Dissociation Constants of Inorganic Acids and Bases, Perrin, D. D. (1969), Butterworths, London.

Ionisation Constants of Organic Acids in Aqueous Solution, Serjeant, E. P. and Dempsey, B. (1979), Pergamon, Oxford.

Among more limited collections there are:

'Heats of proton ionization and related thermodynamic quantities', Izatt, R. M. and Christensen, J. J. (1968), in *Handbook of Biochemistry* (ed. Sober, M. A.), J-49, Chemical Rubber Company, Cleveland, Ohio.

'Ionization constants of acids and bases', Jencks, W. P. and Regenstein, J. (1968), in *Handbook of Biochemistry* (ed. Sober, M. A.), J-150, Chemical Rubber Company, Cleveland, Ohio.

1.4 'Thermodynamic' and 'practical' ionization constants

Equation (1.6), where only activities are involved, gives the 'thermodynamic' pK_a. When concentrations of the acid and base species are used, a 'practical' or 'mixed' constant, pK'_a, is given:

$$pK'_a = (H_3O^+)[B]/[A] \qquad (1.10)$$

where square brackets denote concentrations in moles l^{-1}.

A relation between pK_a and pK'_a can be derived beginning from

$$a_i = c_i f_i \qquad (1.11)$$

where f_i is the activity coefficient (molar scale) of an ion of activity, a_i, and molar concentration, c_i. For an ion of charge, z_i, the activity coefficient is given for dilute solutions by the extended Debye–Hückel equation which is commonly written in the form (Davies 1938)

$$-\log f_i = Az_i^2 I^{\frac{1}{2}}/(1 + I^{\frac{1}{2}}) - 0.1 z_i^2 I \qquad (1.12)$$

where the ionic strength

$$I = \tfrac{1}{2}\Sigma(c_i z_i^2) \qquad (1.13)$$

and A is a Debye–Hückel parameter, values of which are given in Table 1.3. Equation (1.12) is a good approximation at ionic strengths

Table 1.3 *Values of the Debye–Hückel constant, A, for aqueous solutions*

$t°$ C	A	$t°$ C	A	$t°$ C	A
0	0.4918	30	0.5161	70	0.5625
10	0.4989	40	0.5262	80	0.5767
20	0.5070	50	0.5373	90	0.5920
25	0.5115	60	0.5494	100	0.6086

below 0.1. Most of the reliable ionization constants that have been determined experimentally have been extrapolated to zero ionic strength giving 'thermodynamic' values.

Applying Equations (1.6), (1.10)–(1.13) to the systems

$$HA^{(n-1)-} \rightleftharpoons H^+ + A^{n-} \qquad (1.14)$$

and

$$HB^{(n+1)+} \rightleftharpoons H^+ + B^{n+1} \qquad (1.15)$$

gives respectively the relations

$$pK'_a = pK_a - (2n-1)AI^{\frac{1}{2}}/(1 + I^{\frac{1}{2}}) + 0.1(2n-1)I \qquad (1.16)$$

and

$$pK'_a = pK_a + (2n+1)AI^{\frac{1}{2}}/(1 + I^{\frac{1}{2}}) - 0.1(2n+1)I \qquad (1.17)$$

pK' values are useful in calculating buffer concentrations and in kinetic studies. These calculations are good approximations, at least up to an ionic strength of 0.1, but diverge progressively at higher ionic strengths. As an example, the pK'_a of acetic acid is less than pK_a by 0.02, 0.05, 0.11 and 0.14, respectively, for ionic strengths of 0.001, 0.01, 0.1 and 0.2, while the pK'_a of ammonium ion is increased by the same amounts.

1.5 Temperature effects

From the defintion of $\Delta S^0 = -d(\Delta G^0)/dT$ and $-\Delta G^0 = RT \ln K_a$, where T is in degree Kelvin (K), it follows that for ΔS^0 in J deg^{-1} $mole^{-1}$

$$-d(pK_a)/dT = (pK_a + 0.052\Delta S^0)/T \tag{1.18}$$

If the approximation is made that the change in ΔS^0 with temperature is negligible it is possible to calculate pK_a values at temperatures other than 298 K.

For simple carboxylic acids, pK_a is around 4–5, and ΔS^0 is in the region -88 ± 17 J deg^{-1} $mole^{-1}$ at 25° so the term in parentheses in Equation (1.18) is approximately zero. Examples are:

	pK_a	ΔS_0	$-d(pK_a)/dT$
acetic acid	4.76	−73.6	0.0031
butanoic acid	4.82	−102	−0.0016
benzoic acid	4.20	−79.1	0.0003

Thus the pK_a values of common carboxylic acids vary only slightly with ambient temperature.

On the other hand, with phenols the greater pK_a values offset the effect of the entropy change (-100 ± 17 J deg^{-1} $mole^{-1}$ at 25°) on $d(pK_a)/dT$, so the pK_a falls slightly with temperature rise. Typical examples are phenol and its 3-chloro-, 2-hydroxymethyl-, 4-nitro- and 2,5-dimethyl derivatives, for which $d(pK_a)/dT = -0.013 \pm 0.002$.

The entropy change for proton loss from heterocyclic compounds such as benzimidazole and purine is also around −105 entropy units. Insertion of this value into Equation (1.18) leads to

$$-d(pK_a)/dT = (pK_a - 5.4)/T \tag{1.19}$$

Recently this equation was reported to give pK_a values extrapolated from 25° to 85°C that agreed in general within 0.3 pK units with values obtained by isotope exchange measurements (Jones and Taylor, 1980).

The entropy change for the reaction

$$BH^+ \rightleftharpoons B + H^+$$

is less than for carboxylic acids and phenols because the number of ions and their charges do not change. For proton loss from monovalent organic cations, ΔS^0 lies in the range -17 ± 25 J deg^{-1} mole^{-1} giving

$$-\mathrm{d}(\mathrm{p}K_a)/\mathrm{d}T = \{(\mathrm{p}K_a - 0.9)/T\} \pm 0.004 \quad (1.20)$$

so that over a wide range of pK_a values the temperature coefficient varies linearly with pK_a. Experimental values extend from -0.003 for 2-nitroaniline (pK_a = -0.26) to 0.035 for pyrrolidine (pK_a = 11.31). A table summarizing the observed and predicted effects of temperature on pK_a values of 54 organic bases is given elsewhere (Perrin, 1964). The predicted decreases in pK_a with rise in temperature for conjugate acids, BH^+, are given in Table 1.4.

Table 1.4 *Predicted effect of temperature change on* pK_a *values of the conjugate acids,* BH^+

pK_a	$-\mathrm{d}(\mathrm{p}K_a)/\mathrm{d}T$	pK_a	$-\mathrm{d}(\mathrm{p}K_a)/\mathrm{d}T$
1.0	0.000	8.0	0.024
2.0	0.004	9.0	0.027
3.0	0.007	10.0	0.031
4.0	0.010	11.0	0.034
5.0	0.014	12.0	0.037
6.0	0.017	13.0	0.041
7.0	0.020	14.0	0.044

Similarly, the relation

$$-\mathrm{d}(\mathrm{p}K_a)/\mathrm{d}T = (\mathrm{p}K_a)/T \quad (1.21)$$

gives a good approximation in the reaction

$$BH^{2+} \rightleftharpoons BH^+ + H^+ \quad (1.22)$$

Values of $-\mathrm{d}(\mathrm{p}K_a)/\mathrm{d}T$ vary from 0.021 for 1,2-diaminocyclohexane (pK_a = 6.34) to 0.033 for 1,6-diaminohexane (pK_a = 9.83).

The basic pK_a values (NH_3^+) of amino acids and zwitterions have temperature coefficients similar to those of aminium ions, ranging from -0.006 for aniline-2-sulphonic acid ($pK_a = 2.46$) to -0.025 for proline ($pK_a = 10.64$), and the same formula as for monovalent cations (Equation (1.20)) serves for their prediction.

1.6 Solvent effects

With acids ionizing according to

$$HA \rightleftharpoons H^+ + A^-$$

two ions are generated for each neutral molecule that dissociates. This equilibrium is very sensitive to the dielectric constant of the medium, so that the pK_a increases markedly (the acid becomes weaker) in solvents of low dielectric constant.

On the other hand, the ionization

$$BH^+ \rightleftharpoons B + H^+$$

involves no change in the number of ions, so that the effect of solvent is much less (a slight decrease in pK_a with decreasing dielectric constant). For this type of equilibrium, measurements made in mixed solvents can probably be extrapolated to give values in aqueous solution.

1.7 Experimental determination of pK_a values

Notwithstanding the thousands of pK_a values available in the literature, this information is frequently lacking for particular acids and bases of interest. When time and materials permit, it is desirable to determine unknown constants experimentally.

The methods used for the determination of ionization constants in solution usually depend upon a quantitative assessment of the ratio of deprotonated/protonated forms for a compound under known conditions of acidity for the solvent medium. Although it is often possible to determine this ratio with high precision, the reliability of any of these methods depends ultimately upon the accuracy with which the appropriate acidity function can either be measured or assigned. The only acidity function of thermodynamic significance in aqueous solution is the quantity $p(a_H \gamma_{Cl})$ which can be measured potentiometrically with the cell

$$Pt;H_2|H^+ \text{ in aqueous solution, } Cl^- \text{(known molality)}|AgCl;Ag$$

However, the scale most frequently used for the determination of ionization constants is based on the more conveniently measured quantity, pH, a quasi-thermodynamic derivative of $p(a_H \gamma_{Cl})$ (Bates, 1973). As commonly measured in a cell with liquid junction containing a glass electrode and a calomel reference half-cell, the range of utility for these determinations extends from pH 2 to pH 12. Outside this range, the accuracy of pH measurements becomes uncertain as a result of variability both in the magnitude of the liquid junction potential and also in the response of the glass electrode. Acidity scales are available for extending into the strongly acid or alkaline regions the range of ionization constants that can be determined, but a given scale is only appropriate for a particular class of compound (Rochester 1970). Failure to select the correct type of scale can lead to large errors in the derived value of the ionization constant that will be independent of the precision of the method by which the ratio deprotonated form/protonated form has been determined. Two general methods are available for the determination of this ratio. These are based upon:

1. a knowledge of the stoicheiometry of the solution for which either $p(a_H \gamma_{Cl})$ or pH is the measured quantity;
2. a quantitative analysis of each of the two components in a solution of known acidity function.

The most accurate of the stoicheiometric methods uses the type of cell given above (King, 1965). The quantity $p(a_H \gamma_{Cl})$ is measured for a series of solutions each of which, for example, contains the same weights of pure acid and its pure sodium salt, but a different and known concentration of chloride ion. Sometimes the ratio can be obtained directly from the amount of each component weighed, but more usually the ratio must be calculated to allow for hydrolysis of one of the species. In the more convenient but less accurate potentiometric titration method, pH is the measured quantity and the ratio is calculated from the stoicheiometry deduced from the weight of pure compound taken and the known volumes of standardized titrant added (Albert and Serjeant, 1971). Usually the solubility of the compound needs to be greater than 10^{-3}M for the application of these methods.

The analytical methods for the determination of the ratio deprotonated form/protonated form depend upon some difference in property or response that allows a quantitative discrimination to be made between the two forms of the compound. For example, if a study of the spectrum (absorbance versus wavelength) of the pure deprotonated form obtained in a solution of high pH reveals significantly different features from the spectrum of the pure protonated form obtained at a lower pH, then it is possible to deduce the ratio by measuring the absorbances of solutions at values of pH that are intermediate between the upper and lower limits. The method is particularly applicable to sparingly soluble aromatic and hetero-aromatic acids and bases having solubilities less than 10^{-3}M. Accurate values of ionization constants can be obtained if solutions are prepared in buffers having known $p(a_H \gamma_{Cl})$ values (Bolton, Hall and Reece, 1966); less accurate values are obtained with pH buffer solutions and very approximate values if other types of acidity functions are used (Albert and Serjeant, 1971). The other methods that come into this general category have been reviewed (Cookson, 1974). The pK values of sparingly soluble compounds that do not exhibit suitable spectra can sometimes be determined by solubility or distribution studies. Problems associated with high acidity or alka-linity outside the range pH 2–12 are such that pK_a values derived from these methods are also unreliable if they are outside this range. For details of methods and calculations the reader is referred to Albert and Serjeant (1971).

On the other hand it would often be helpful to have an approximate pK_a value, correct to within a few tenths of a pH unit, and the present monograph is addressed to this purpose.

Chapter Two

Molecular Factors that Modify pK_a Values

Qualitatively, factors that modify pK_a values are well understood. They comprise mainly inductive, electrostatic and electron-delocalization (mesomeric) effects, together with contributions from hydrogen bonding, conformational differences and steric factors. Where a substituent tends to stabilize a cation more than the corresponding neutral molecule, or a neutral molecule more than the anion derived from it, the pK_a of the acid will be raised. That is, the substituent will be acid-weakening or base-strengthening. Conversely, increased stabilization of an anion relative to the neutral species is acid-strengthening or base-weakening. Consideration of factors that modify the relative stabilities of these species can thus provide insights as to why pK_a values vary but they do not permit calculations of the differences involved. It is nonetheless, useful to have even a qualitative indication of how substituents affect pK_a values.

2.1 Inductive and electrostatic effects

Electrical work is required to remove a proton from an acidic centre and transfer it to a solvent molecule, or, conversely, to remove a proton from a solvent molecule and transfer it to an organic base. The amount of work done is influenced by the locations and distributions of the dipoles and electrical charges. The inductive effects are transmitted through bonds in the molecule, in contradistinction to the electrostatic effects which are thought to operate across the low-dielectric cavity provided by the solute or through the solvent. Inductive effects would be expected to remain approximately constant in *cis* and *trans* isomers, the major difference in pK_a values arising from differences in the electrostatic field effects, but because inductive and electrostatic effects usually operate in similar directions it is not usually possible to separate them.

A substituent is considered to have a $+I$ effect (acid-weakening) if its insertion into a molecule in place of a hydrogen atom increases the electron density at other points of the molecule. If, instead, it decreases the electron density it has a $-I$ effect and is acid-strengthening or base-weakening. In chloroacetic acid, the chlorine atom of the carbon–chlorine bond is electron withdrawing with respect to carbon, so that the electrons forming the C–Cl bond are displaced towards the chlorine. This makes the methylene carbon more electron deficient than in the corresponding methyl group of acetic acid. In turn, the carboxyl oxygens are more electron deficient, thus increasing the ease with which a proton can be removed from the free acid.

Inductive effects fall off rapidly with distance in saturated hydrocarbons, so that the effect of a group $-CH_2R$ is only about two fifths as great as for the group $-R$. Attenuation is less rapid in unsaturated systems such as $-C=C-$, $-C \equiv C-$ or $-C=N-$. The effect of group such as $-C=C-$ is comparable with $-CH_2-$. According to Jaffe (1953) the transmission factor for $-CH=CH-$ is 0.51 and for (benzene ring) it is 0.30 (compared to 0.41 for $-CH_2-$). Replacing a $-CH_2-$ by $-O-$, $-S-$, $-Se-$, $-SO-$, $-SO_2-$, or $-NH-$ group has little effect on the ability of chains to transmit polar effects between substituents and reaction centres.

2.2 Mesomeric effects

Mesomeric effects ($\pm M$) arise from π-electron delocalization ('resonance'). They contribute appreciably to the ease with which the strength of an acid or a base is modified by remote substituents, especially in aromatic or heteroaromatic systems that bear *ortho* or *para* substituents (the mesomeric effects of *meta* substituents are negligible). Mesomeric effects may enhance or oppose inductive effects, as shown in Table 2.1 in which common substituents are grouped in terms of these effects.

From the relation

$$\Delta G^0 = 2.303RT.\mathrm{p}K_a \qquad (2.1)$$

a change of 5.7 kJ mole^{-1} in ΔG^0 produces, at 25°C, a change of 1 pH unit in the value of pK_a. Free energy differences of several kilojoules

Table 2.1 *Inductive and resonance effects of substituents*

$+I$ (acid-weakening)	$-CO_2^-$, $-O^-$, $-NH^-$, $-$alkyl
$-I$ (acid-strengthening)†	$-NH_3^+$, $-NR_3^+$, $-NO_2$, $-SO_2R$, $-CN$, $-F$, $-Cl$, $-Br$, $-I$, $-CF_3$, $-COOH$, $-CONH_2$, $-COOR$, $-CHO$, $-COR$, $-OR$, $-SR$, $-NH_2$, $-C_6H_5$
$+M$ (acid-weakening)	$-F$, $-Cl$, $-Br$, $-I$, $-OH$, $-OR$, $-NH_2$, $-NR_2$, $-NHCOR$, $-O^-$, $-NH^-$, $-$alkyl
$-M$ (acid-strengthening)	$-NO_2$, $-CN$, $-CO_2H$, $-CO_2R$, $-CONH_2$, $-C_6H_5$, $-COR$, $-SO_2R$

† Approximately in decreasing order.

can result from charge delocalization so that acids or bases may, in such cases, be appreciably stronger or weaker than would otherwise be expected. Examples include the guanidinium cation (*2.1*), the amidinium cation (*2.2*), imidazolium ion (*2.3*), and partially reduced hereroaromatic systems such as 1,4,5,6-tetrahydropyrimidinium ion (*2.4*), all of which are conjugate acids of strong bases. A feature common to all these examples is that they contain an imide nitrogen *alpha* to an $-NH-$ group. The cations are stabilized by amidinium-type resonance.

Conversely, the *p*-nitrophenate ion (*2.5*) is stabilized by resonance so that *p*-nitrophenol is a stronger acid than would otherwise be expected. The anomalously high pK_a (pK_a = 12.34) of 1,8-bis(dimethylamino)naphthalene is attributed to steric interaction of

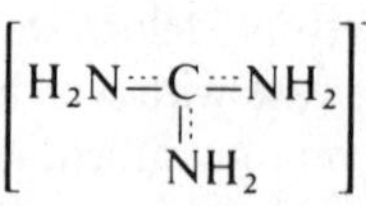

$[H_2N{\cdots}C{\cdots}NH_2]^+$ with NH_2 on C (*2.1*)

$[H_2N{\cdots}CH{\cdots}NH_2]^+$ (*2.2*)

$[$NH, N–H imidazolium ring$]^+$ (*2.3*)

$[$NH, N–H tetrahydropyrimidinium ring$]^+$ (*2.4*)

$[$O–C₆H₄–NO₂ ~ O=C₆H₄=NO₂ ~ O=C₆H₄=NO₂$]^-$ (*2.5*)

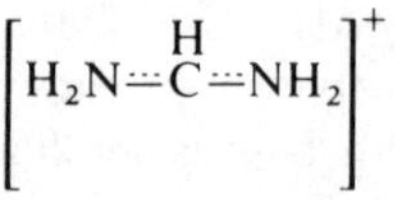

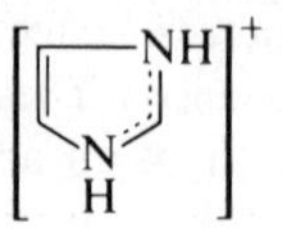

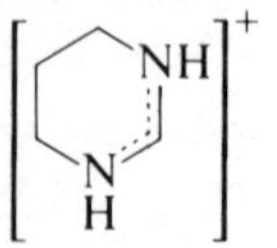

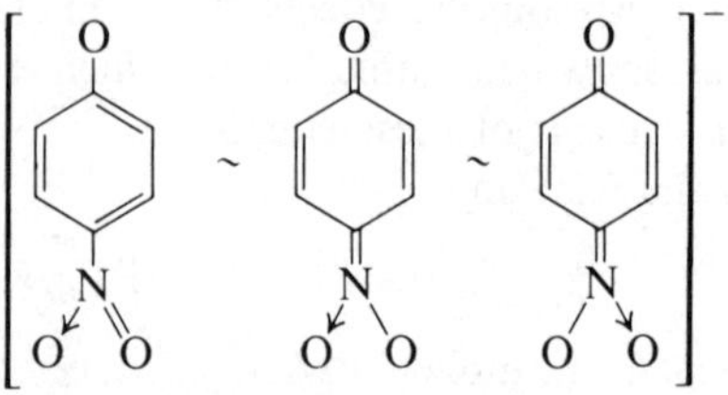

the methyl groups preventing the lone pairs of electrons on the nitrogen atoms from lying in the plane of the naphthalene molecule, leading to steric inhibition of resonance stabilization.

2.3 Steric effects

Steric restraints imposed by the double bond in *cis*- and *trans*-isomers of diamines and dicarboxylic acids locate the acidic and basic centres at different distances from one another so that the pK_a values of fumaric acid (3.10, 4.60) differ from its *cis*- isomer, maleic acid (1.91, 6.33). Such differences are common in pairs of geometrical or conformational isomers and may be due to several factors. Thus there may be overlapping of solvation shells around the reaction centres in the *cis*- isomers or the effect may be due to electrostatic repulsion arising from the closer proximity of the reaction centres. This leads to the second pK_a of a *cis*- diamine being less than that for the *trans*-isomer, and hence to the *cis*- isomer being the weaker base. Similarly, the second pK_a of maleic acid indicates that it is a weaker acid than fumaric acid. On the other hand, internal hydrogen bonding is also possible and, in the monoanion of maleic acid, may be expected to stabilize this species and as a consequence to weaken the second dissociation step.

Carboxyl groups are larger than hydroxyl or primary amino groups, and hence are more susceptible to steric interactions. Primary steric hindrance to carboxyl anion formation may explain why pivalic acid (2,2-dimethylpropanoic acid, pK_a = 5.04) is a weaker acid than pentanoic acid (pK_a = 4.76) or 2-methylbutanoic acid (pK_a = 4.76) whereas the same substitution makes little difference to the pK_a values of the corresponding primary amines (the pK_a of 2-amino-2-methylpropane is 10.68 at 25°, pK_a of 1-amino-butane is 10.61). That this is a matter of degree rather than of kind is seen if slightly more complex compounds are considered. Location of the alkyl groups around the nitrogen atom of 3-amino-2,4-dimethylpentane increases the difficulty of protonating the amino group, so that this base is weaker than expected. The pK_a of 3-amino-2,2,4,4-tetramethyl-pentane has not been recorded but would be expected to be weaker still.

Crowding by the two methyl groups of 3,5-dimethyl-4-nitrophenol prevents the nitro group from achieving planarity with the ring, thereby decreasing resonance stabilization of the anion. The measured pK_a for the dimethyl compound is about 1 pH unit weaker than expected from the parent compound. Thus:

pK_a of phenol = 10.00
pK_a of 4-nitrophenol = 7.16
Deerease in pK_a = 2.84
pK_a of 3,5-dimethylphenol = 10.19
Expected pK_a of 3,5-dimethyl-4-nitrophenol = 10.19 − 2.84 = 7.35
Observed pK_a = 8.25

Steric hindrance to solvation on cation formation by peri- and meso-substituted aromatic amines such as 9-anthrylamine and 1-aminotriphenylene causes them to be much weaker bases than aniline.

Intramolecular hydrogen bonding can also be important, as can be seen by comparing the pK_a (1.3) of 2,6-dihydroxybenzoic acid (where this bonding probably occurs) with the pK_a (4.47) of 3,5-dihyroxybenzoic acid (where it does not). Likewise, salicylic acid is stronger than 2-methoxy-, or 3- or 4-hydroxy- benzoic acids.

Conformational effects influence the pK_a values of cyclohexanecarboxylic acids and related alicylic acids, axial carboxyl groups being weaker than equatorial ones. A relation has been deduced and used in elucidating the stereochemistry of *cis* and *trans* decalincarboxylic acids and resin acids (Sommer *et al.*, 1963). For a discussion of these and other applications to structure determination, see Barlin and Perrin (1972).

2.4 Statistical factors

When a polybasic acid has n groups, each of which has an equal probability of losing a proton, the observed pK_a will be less by $\log n$ than the pK_a of a closely related monobasic acid. This 'statistical effect' arises because there are n equivalent ways of losing a proton but only one site to which the proton can be restored. Similarly, for the second proton loss, the correction becomes $\log[(n-1)/2]$, then $\log[(n-2)/3]$, and so on. Thus, for a molecule such as butanedioic acid, which has two identical acidic groups, loss of a proton from either group leads to the same monoanion. The consequence is that the first ionization constant, K_{a1}, for the dibasic acid is twice as large as that for the closely related monobasic acid, that is, the observed pK_{a1} is 0.3 (= log 2) units less than would be expected from a consideration of factors other than probability. Conversely, the monoanion has only one ionizable proton whereas the dianion has two identical sites for proton addition, so that the second ionization step, K_{a2}, appears to be weaker by a factor of two, and the observed pK_{a2} to be greater by 0.3 than anticipated. Similarly, for a base with n basic centres, the measured pK_a of greatest magnitude, pK_{an}, will be greater than anticipated by $\log n$, and so on.

When predicting pK_a values, statistical corrections must be applied *after* the molecular effects have been taken into account. Conversely, in the determination of sigma values from Hammett and Taft equations (see Chapter 5), statistical corrections, where appropriate, must be made before calculations are undertaken.

2.5 Microscopic constants

Where the species, RH_2, has two pK_a values that are of comparable magnitude, the dissociation of one group is affected by the other, and the following equilibria apply:

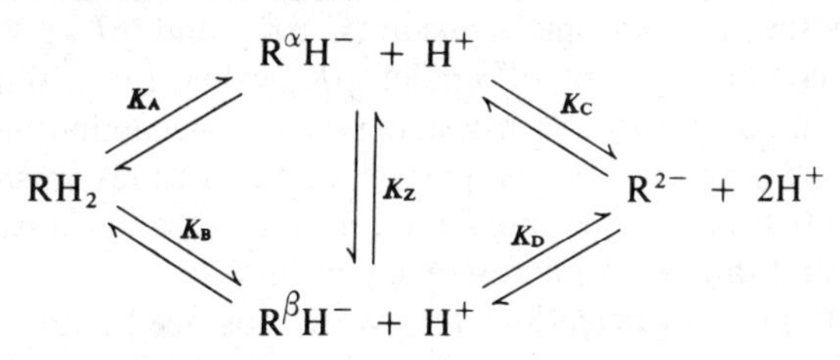

K_A, K_B, K_C and K_D are 'microscopic' constants; and K_Z is the tautomeric ratio. The experimentally measured (or 'macroscopic') constants, K_1 and K_2, refer, respectively, to the equilibria:

$$RH_2 \rightleftharpoons R^{\alpha}H^- \quad \text{and} \quad R^{\beta}H^- + H^+ \tag{2.2}$$

and

$$R^{\alpha}H^- \quad \text{and} \quad R^{\beta}H^- \rightleftharpoons R^{2-} + H^+ \tag{2.3}$$

The various constants are related as follows:

$$K_1 = K_A + K_B \tag{2.4}$$

$$\frac{1}{K_2} = \frac{1}{K_C} + \frac{1}{K_D} \tag{2.5}$$

$$K_Z = \frac{K_A}{K_B} = \frac{K_D}{K_C} \tag{2.6}$$

When K_Z approaches unity, the measured 'macroscopic' constants are a composite of the processes actually occurring, but when K_Z is large, pK_1 approximates to pK_a. When K_Z is small, pK_1 approximates to pK_B.

An example of where these equations apply is in zwitterion formation as shown:

$$\overset{+}{H_3N}-R-COO^- + H^+$$

$$\overset{+}{H_3N}-R-COOH \underset{K_B}{\overset{K_A}{\rightleftharpoons}} \quad \updownarrow K_Z \quad H_2N-R-COO^- + 2H^+ \quad (K_C,\ K_D)$$

$$H_2N-R-COOH + H^+$$

With α-aminoacids, equilibria are so heavily in favour of the zwitterionic pathway involving K_A and K_C that negligible error is introduced if the pK_a values predicted for pK_A and pK_B are taken as those for the 'macroscopic' constants, pK_1 and pK_2. Where a molecule contains groups of similar pK_a value, for example, the mercapto and amino groups in cystein (and in other aminothiols), the observed pK_a values are composites of the various microscopic constants. Methods of pK_a prediction must be based on a particular structure and thus yield microscopic constants.

For examples of calculations for zwitterions, see Section 7.3.

2.6 Tautomerism

Many heterocyclic systems have an inherent tendency to tautomerize or to form zwitterions. Keto–enol, lactam–lactim and thiol–thione tautomerism are all examples in which a proton migrates within the molecule. Similarly, in amphoteric molecules where the pK_a of the basic group is greater than, or near to, the pK_a of the acidic group, zwitterion formation is possible. In all cases, the pK_a values of the tautomeric forms may be very different from each other.

2.7 Structure and ionization

Substituents in saturated hydrocarbon rings and chains produce only inductive and field effects, the magnitude of which depends in part on the distance which separates the substituents from the acidic or basic centre. Thus because $-Cl$ has a $-I$ effect it is to be expected that

$$\text{p}K_a \text{ of } ClCH_2COOH < \text{p}K_a \text{ of } ClCH_2CH_2COOH < \text{p}K_a \text{ of } ClCH_2CH_2CH_2COOH$$

Substituent effects are approximately additive so

$$\text{p}K_a \text{ of } CH_3COOH > \text{p}K_a \text{ of } ClCH_2COOH > \text{p}K_a \text{ of } Cl_2CHCOOH$$

Aromatic acids and bases are influenced by mesomeric effects: an electronegative substituent withdraws charge from the rest of a π-electron system, but only from carbon atoms separated from it by an odd number of bonds. This gives an alternating pattern of 'normal' and electron deficient carbon atoms, so that an *ortho* or *para* substituent exerts both inductive and mesomeric effects, whereas a *meta* substituent exerts only an inductive effect.

Six-membered heteroaromatic rings containing nitrogen are a major class of organic acids and bases. They comprise pyridines, pyrimidines, pyridazines, pyrazines, triazines and their benzologues such as quinolines, cinnolines and pteridines. Annelation of benzene rings on to a pyridine or its derivatives makes only a small change in acid or basic strength. Bonding of the pyridine nitrogen involves planar trigonal (sp^2) orbitals so that this amine is intermediate in basicity between an aliphatic amine (sp^3) and a cyano compound (*s* orbital). Insertion of another nitrogen into the ring lowers the basic strength ($-I$, $-M$ effects) to about the same extent as for a nitro group, except where the first nitrogen is part of an amide group. In this case, the second nitrogen serves as an independent basic centre and basic strength may increase. A methoxy group has a $-I$, $+M$ effect so that it is base-weakening when it is substituted in the *beta* position (which has no M effect) and base-strengthening ($|+M| > |-I|$) when it is an *alpha* or *gamma* substituent.

When an amino group is attached to a nitrogen-containing aromatic heterocycle, protonation occurs firstly on the aromatic ring nitrogen. However, when an aminoalkyl group is attached to such a ring, protonation occurs firstly on the non-ring nitrogen. The neutral molecule of 4-aminopyridine is stabilized by benzene-type resonance but the imino form of the cation has even greater resonance stabilization so that the base strength of 4-aminopyridine is increased. The N-methyl analogue, 1,4-dihydro-l-methyl-4-iminopyridine, does not have to lose the resonance stabilization of the neutral molecule so its basic strength (pK_a = 12.5) is even greater. Explanations in terms of resonance stabilization have been given for the basic strengths of 2-aminoquinoline being stronger than 2-aminopyridine, 1-aminoisoquinoline being stronger than 3-aminoisoquinoline, and 7-aminoquinoline being stronger than 5-aminoquinoline (Gore and Phillips, 1949).

Zwitterion formation is common if amino and carboxyl groups are present on the same ring. Hydroxyl and thiol groups α- or γ- to a ring nitrogen tautomerize to dihydro-oxo and thio- derivatives. Where the possibility exists for the formation of *ortho* or *para* quinonoid structures, as in pyrimidines, quinazolines or pteridines, the *ortho* quinonoid isomer is the more stable.

Five-membered heterocyclic rings usually contain $-NH-$, $-O-$, or $-S-$. The nitrogen compounds are mainly pyrroles, imidazoles and purines, and are weakly acidic because of ionization of the proton on the ring nitrogen. Imidazole (pK_a = 6.95) is a much stronger base than pyrrole (p$K_a \sim -0.3$) owing to resonance stabilization of the cation:

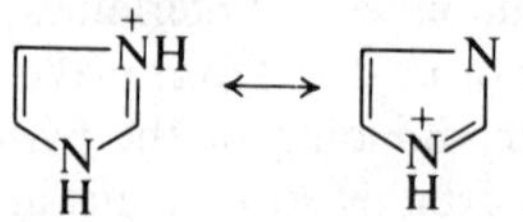

Hammett values (see Section A.3) indicate that a nitrogen in the imidazole ring in purines is protonated preferentially before the nitrogens in the pyrimidine ring.

Chapter Three

Methods of pK_a Prediction

The extensive collections of pK_a data currently available make it possible, without much difficulty, to predict from classes of compound and by analogy the pK_a values of most acids or bases to within a few tenths of a pH unit.

Our approach is strictly pragmatic and hence is more concerned with experimentally derived linear free energy relations for pK_a prediction than with attempts to derive pK_a values from first principles.

3.1 Linear free energy relationships

At a given temperature, the pK_a value for the ionization of a proton from a molecule is directly related to the change in free energy for the reaction:

$$\Delta G^0 = 2.3026\, RT \,.\, \mathrm{p}K_a \qquad (3.1)$$

The effects of substituents on pK_a value can therefore be discussed in terms of factors which affect free energies. The change in ΔG that is produced by inserting a substituent leads to a corresponding change in pK_a. Where these changes are not very great they are found to be approximately additive. This observation provides the basis for the Hammett and Taft equations which are the most widely used of the methods of prediction. These methods are discussed in Chapters 4 and 5.

3.2 Prediction from class of compound

Changes in the length or extent of branching of an aliphatic hydrocarbon usually have little effect on the pK_a of the corresponding amine or carboxylic acid in aqueous solution. This is because the inductive effect of an alkyl group is small and falls off rapidly with distance from the reaction centre, so that only atoms which are near neighbours to the acidic or basic centres exert a significant effect on the observed pK_a values.

3.2.1 Aliphatic amines

Thus the pK_a of 1-aminododecane (10.60 at 25° and $I = 0.01$–0.1) is only marginally less than the value of 10.70 for ethylamine or 10.63 for methylamine. From Table 3.1, which lists representative pK_a values, it is seen that most pK_a values for unsubstituted aliphatic amines at 25°

Table 3.1 pK_a *Values of aliphatic amines at 25° and* $I = 0.01$–0.1

Primary amine	pK_a	
CH_3NH_2	10.63	
$CH_3CH_2NH_2$	10.70	
$CH_3[CH_2]_2NH_2$	10.60	
$CH_3CH_2CH(CH_3)NH_2$	10.56	
$(CH_3CH_2)_2CHNH_2$	10.42	
$CH_3[CH_2]_2CH(CH_3)NH_2$	10.67	
$CH_3CH_2C(CH_3)_2NH_2$	10.72	
$(CH_3CHCH_3)_2CHNH_2$	10.23	
$CH_3[CH_2]_4C(CH_3)_2NH_2$	10.56	
$CH_3[CH_2]_6NH_2$	10.67	
$(CH_3)_3CNH_2$	10.68	
$(CH_3CH_2)_2C(CH_3)NH_2$	10.63	
$(CH_3CH_2)_3CNH_2$	10.59	
Secondary amine	**pK_a**	**$11.1 - 0.2 \times n$†**
$(CH_3)_2NH$	10.78	10.7
$(CH_3CH_2)_2NH$	11.02	11.1
$(CH_3[CH_2]_2)_2NH$	11.00	11.1
$(CH_3CHCH_3)_2NH$	11.05	11.1
$(CH_3[CH_2]_3)_2NH$	11.25	11.1
$(CH_3[CH_2]_5)_2NH$	11.01	11.1
$(CH_3[CH_2]_7)_2NH$	11.01	11.1
$(CH_3[CH_2]_6)(CH_3)NH$	10.82	10.9
Tertiary amine	**pK_a**	**$10.5 - 0.2 \times n$‡**
$(CH_3)_3N$	9.80	9.9
$(CH_3CH_2)_3N$	10.75	10.5
$(CH_3CH_2)N(CH_3)_2$	10.05	10.1
$(CH_3CH_2)_2N(CH_3)$	10.35	10.3
$(CH_3[CH_2]_2)N(CH_3)_2$	10.05	10.1
$(CH_3CHCH_3)N(CH_3)_2$	10.36	10.1
$(C(CH_3)_3)(CH_3)_2N$	9.97	10.1
$(CH_3[CH_2]_3)(CH_3)_2N$	10.08	10.1
$(CH_3CH_2CHCH_3)(CH_3)_2N$	10.46	10.1
$(CH_3[CH_2]_2)_3N$	10.66	10.5

†n = the number of methyl groups bound to the basic nitrogen atom. See Equation (3.3).
‡ See Equation (3.4).

and $I = 0.01–0.1$ fall into the following ranges:

$$\text{primary amines,} \quad pK_a = 10.6 \pm 0.2 \tag{3.2}$$

$$\text{secondary amines,} \quad pK_a = 11.1 \pm 0.1 - n \times 0.2 \tag{3.3}$$

$$\text{tertiary amines,} \quad pK_a = 10.5 \pm 0.2 - n \times 0.2 \tag{3.4}$$

where n is the number of methyl groups bound to the basic nitrogen atom.

The decrease in basic pK_a in going from secondary to tertiary amines has been suggested to be due to the cation of the latter having only one hydrogen atom bound to the nitrogen atom, so that stabilization by hydrogen bonding of the type $\geqslant N^+H \ldots OH_2$ is much less effective than for the secondary amine.

Thus, pK_a values of aliphatic amines are not appreciably changed if large alkyl groups or chains are replaced by methyl or ethyl groups.

3.2.2 Aliphatic carboxylic acids

The pK_a values (at 25° and $I = 0$) for aliphatic acids also show very little variation with chain length. Relations are:

$$\text{for } RCH_2COOH \text{ or } R_1R_2CHCOOH, \quad pK_a = 4.8 \pm 0.1; \tag{3.5}$$

$$\text{for } R_1R_2R_3COOH, \quad pK_a = 5.0 \pm 0.1. \tag{3.6}$$

Structural simplification can be made for large alkyl groups or chains as for aliphatic amines. Table 3.2 illustrates the generality of the relations (3.5) and (3.6).

3.3 Prediction based on analogy

The simplification that alkyl rings or chains can be replaced by methyl or ethyl groups in assessing their effect on pK_a values extends to alicyclic rings and saturated carbocyclic rings fused to aromatic and heterocyclic rings (provided the former are essentially strain free). Where there is ring formation through a basic nitrogen the pK_a is raised by 0.2 pH units for one ring or 0.3 for two rings relative to the open chain alkylamines. Examples are given in Table 3.3. Cyclic amines such as pyrrolidine (pK_a = 11.2) have pK_a values similar to that of diethylamine (pK_a = 11.1) if allowance is made for the extra 0.2 pH unit to be added for ring closure. Similarly, N-alkyl polymethyleneimines have pK_a values close to those of aliphatic tertiary

Table 3.2 pK_a *Values of aliphatic acids at 25° and I = 0*

RCH_2COOH	pK_a
CH_3COOH	4.76
CH_3CH_2COOH	4.86
$CH_3[CH_2]_2COOH$	4.83
$CH_3[CH_2]_3COOH$	4.84
$CH_3[CH_2]_4COOH$	4.85
$R_1R_2CHCOOH$	
$(CH_3)_2CHCOOH$	4.88
$C_2H_5CH(CH_3)COOH$	4.80
$CH_3[CH_2]_2CH(CH_3)COOH$	4.79 (18°)
$(C_2H_5)_2CHCOOH$	4.74
$R_1R_2R_3CCOOH$	
$(CH_3)_3COOH$	5.03
$CH_3CH_2C(CH_3)_2COOH$	5.04 (18°)
$CH_3[CH_2]_3C(CH_3)_2COOH$	4.97

amines. Cyclic amidines and guanidines can be compared with the corresponding open-chain amines.

The pK_a values of carboxylic acids are unchanged by alicyclic or carbocyclic ring formation, thus resembling amines, but in this case no correction is needed for methyl groups (probably because of the greater distance of the acidic carbon from the substituents). Thus the pK_a of cyclohexanecarboxylic acid (pK_a = 4.92) compares with that of propanoic acid (pK_a = 4.87).

Another useful approximation is based on the assumption that in saturated heterocyclic rings the effects operate along all chains in the molecule. Taking morpholine as an example, its pK_a approximates to that of the diether, $CH_3OCH_2CH_2NHCH_2CH_2OCH_3$, plus a small increment (0.2 pK units) for ring formation around the nitrogen.

3.4 Extrapolation

Provided that solvation effects do not vary too widely, the pK_a values of organic bases in different solvents would be expected to fall in the same order as their pK_a values in water. Titration of organic bases in glacial acetic acid afforded estimates of their pK_a values in water within about $\pm$ 0.1 pH unit (Hall, 1930). Mixed solvents are usually unsatisfactory for this purpose because differential solvation of the solute leads to differences in composition between the solvent sheath and the bulk phase. The high dielectric constants of methanol and

Table 3.3 *Prediction of* pK_a *values of cyclic amines from simpler equivalents*

Cyclic amine		Simpler equivalent
NH_2 pK_a = 10.68	cf.	$CH_3CH_2NH_2$ pK_a = 10.65
NH_2 pK_a = 4.42	cf.	NH_2, H_3C, H_3C pK_a = 4.71
NH_2 pK_a = 10.14	cf.	NH_2, C, H_3C, CH_3, CH_3 pK_a = 10.68 (no allowance for strain in aminoadamantine)
H N pK_a = 5.03	cf.	NHR pK_a = 4.85 (for R = CH_3)†‡ = 5.1 (for R = C_2H_5)†
N H pK_a = 11.12	cf.	H_3C CH_3 H_2C CH_2 N H pK_a = 11.02†
N N H pK_a = 13.0	cf.	NH ‖ H_2N-CH pK_a = 12.6
N H pK_a = 11.12	cf. N H pK_a = 11.12	H_3C CH_3 H_2C CH_2 N H pK_a = 11.02†

† A factor of 0.2 pH unit has to be added to each of these compounds when ring formation takes place through the nitrogen.

‡ A factor of 0.2pH unit has to be subtracted for the methyl group bound to the nitrogen.

ethanol make it possible that methanol- or ethanol-water might be used if the pK_a values were plotted against the reciprocal of the dielectric constant of the mixed solvent, provided the dielectric constant of the mixture is kept above about 50 to reduce the possibility of ion-pair formation.

3.5 Predictions based on theory

Attempts to calculate pK_a values from first principles have so far been unsuccessful. The Kirkwood–Westheimer (1938) theory gives ΔpK_a for a charged or a dipolar substituent as:

$$\Delta pK_a = e^2/2.3kTRD_{eff} \tag{3.7}$$

or

$$\Delta pK_a = e\mu\cos\phi/2.3kTR^2D_{eff} \tag{3.8}$$

respectively, where ϕ is the angle between the line joining the centre of the ionizing group to the centre of the dipole and the axis of the dipole. For the other terms, e is the electronic charge, k is the Boltzmann constant, T is the temperature in K, μ is the dipole moment, R is the distance between the two charges, and D_{eff} is the effective dielectric constant.

Calculations are restricted to ellipsoidal molecules with point charges at their foci, or a point charge and a point dipole. Doubt as to what value to assign to D_{eff}, together with approximations involved in the model, prevent calculations from being better than semi-quantitative.

Similarly, there have been quantum mechanical studies of acid–base strength (see, for example, Kuthan *et al.*, 1978, who obtained a correlation between π-electron energy changes during ionization and $\sigma_{m,p}$ and σ_R constants for *m*-and *p*-substituted benzoic acids).

The Born–Haber cycle uses a thermodynamic cycle in which a gas ion is formed, a proton is removed, and the reaction components are solvated. Uncertainty in the free energy changes associated with each step is cumulative and gives too great an error in estimated pK_a. Molecular orbital calculations of the basic strength of nitrogen heterocycles relate to gas phase conditions, and correlations with pK_a values in the aqueous phase are difficult because of differences in solvation energies and variations in steric factors and σ-bond energy changes. For an up-to-date discussion of the solvation of gas ions and the use of such cycles, see Arnett *et al.* (1977) and Reynolds *et al.* (1977).

Chapter Four

Prediction of pK_a Values of Substituted Aliphatic Acids and Bases

4.1 ΔpK_a values for aliphatic acids and amines

Free energy changes produced by inserting substituents into organic molecules are approximately additive. Hence, the simplest way to use a linear free energy relationship for predicting the pK_a of an aliphatic acid or amine where the pK_a of the parent compound is known, is to add to it increments of pK_a (ΔpK_a values) corresponding to the free energy changes produced by inserting the individual substituents. Finally, the total is adjusted by applying statistical factors or other corrections such as a decrease of 0.2 in the pK_a value of an amine for every methyl group bound to the nitrogen, or an increase of 0.2 if the nitrogen atom forms part of a ring.

Where more than one ionization sequence is possible, a pK_a value is calculated for each alternative and the pathway is selected which has protons dissociating from the strongest acidic centre and protons being bound preferentially to the strongest basic centre. The more weakly acid species is favoured where tautomers exist in forms that differ in their acid strengths.

Table 4.1 lists the acid-strengthening ($-\Delta$pK_a) effects of representative substituents attached to the α-carbon atoms of aliphatic acids. These were calculated from data in Table A.1 using

$$-\Delta pK_a = 0.06 + 0.63\,\sigma^* \qquad (4.1)$$

this being the best fit of experimental ΔpK_a versus σ^* values. Similarly, Table 4.2 lists the base-weakening ($-\Delta$pK_a) effects of common substituents attached to β-carbons for aliphatic amines; $-\Delta$pK_a for α-carbons are sometimes anomalous (Clark and Perrin, 1964). The values in Table 4.2 were calculated from data in Table A.1 using

$$-\Delta pK_a = 0.28 + 0.87\,\sigma^* \qquad (4.2)$$

Table 4.1. *Acid-strengthening ($-\Delta pK_a$) effects of substituents commonly attached to α-carbon atoms in aliphatic acids*

Substituent	$-\Delta pK_a$
C_0 —	
$-Br$	1.85
$-Cl$	1.92
$-F$	2.08
$-I$	1.61
$-N_3$	1.71
$-OH$	0.90
$-SH$	1.12
$-NH_2$	0.45
$-NH_3^+$	2.43
$-NO_2$	2.58
$-ONO_2$	2.49
$-SO_3^-$	0.57
C_1	
$-CCl_3$	1.73
$-CF_3$	1.70
$-CN$	2.14
$-COO^-$	−0.61
$-COOH$	1.37
$-CH_2Br$	0.69
$-CH_2Cl$	0.65
$-CH_2F$	0.75
$-CH_2I$	0.69
$-CH_2OH$	0.26
$-OCH_3$	1.20
$-SCH_3$	1.04
$-SCN$	2.22
$-CONH_2$	1.12
$-NHCHO$	1.08
$-NHCONH_2$	0.89
$-SO_2CH_3$	2.38
$-SCONH_2$	1.36
C_2	
$-C\equiv CH$	1.43
$-CH=CH_2$	0.41
$-COCF_3$	2.39
$-CH_2CN$	0.88
$-COCH_3$	1.10
$-OCOCH_3$	1.41
$-COOCH_3$	1.32
$-NH(CH_3)_2^+$	2.81
$-NHCOCH_3$	0.94

Table 4.1 (*Contd.*)

Substituent	$-\Delta pK_a$
C_3	
$-N(CH_3)_3^+$	2.93
$-Si(CH_3)_3$	−0.45
$-NHCOC_2H_5$	1.04
$-NHCOOC_2H_5$	1.31
C_4	
$-CH{=}CHC_2H_5$	0.26
$-CH_2COOC_2H_5$	0.58
C_6	
$-C_6H_5$	0.44
$-OC_6H_5$	1.59
$-SC_6H_5$	1.24
$-O-\text{cyclo}-C_6H_{11}$	1.20
$-S-\text{cyclo}-C_6H_{11}$	1.28
C_7	
$-COC_6H_5$	1.45
$-SCH_2C_6H_5$	1.04
$-CONHC_6H_5$	1.04
C_8	
$-N(COCH_3)C_6H_5$	0.92
C_{10}	
α-naphthyl	0.53
β-naphthyl	0.53
C_n	
$-CH{=}CHR$	0.25
$-OCOR$	~1.7

which was the line of best fit in the plot of experimental ΔpK_a versus σ^* values. The effects are attenuated with increasing chain length, decreasing by a factor of about 0.4 across each $-CH_2-$, that is, the attenuation factor for

$$-CH_2- \sim 0.4 \tag{4.3}$$

A similar additivity of changes in pK_a for the pyridine series enables predictions of the pK_a values of many di- and tri-substituted pyridine derivatives. Table 4.3 gives some of the more commonly found substituents and their $-\Delta pK_a$ effects.

Table 4.2 *Base-weakening ($-\Delta pK_a$) effects of substituents commonly attached to α- and β-carbon atoms in aliphatic amines*†

Substituent	$-\Delta pK_a$	
	α-position‡	β-position§
C_0		
$-Cl$		2.1¶
$-F$		1.9¶
$-OH$		1.1¶
$-SH$		1.3¶
$-NH_2$		0.8
$-NH_3^+$		3.6
$-NO_2$		3.8
$-ONO_2$		3.6
$-SO_3^-$		1.0
C_1		
$-CCl_3$		2.6
$-CF_3$		2.6
$-CN$	5.8	3.2
$-COO^-$	0.8‖, −0.1*	−0.2¶
$-OCH_3$		1.2¶
$-SCH_3$		1.6
$-CONH_2$	2.8	1.7
$-NHCHO$		1.7
$-NHCONH_2$		1.4
$-SO_2CH_3$		3.5
$-SCONH_2$		2.1
C_2		
$-C{\equiv}CH$		1.9¶
$-CH{=}CH_2$		0.8
$-COCF_3$		3.5
$-COCH_3$		1.7
$-COOCH_3$		1.3¶
$-NHCOCH_3$		1.5
C_3		
$-N(CH_3)_3^+$		4.2
$-Si(CH_3)_3$		−0.4
$-NHCOC_2H_5$		1.6
$-NHCOOC_2H_5$		2.0
C_5		
2−pyridyl	2.2	1.14
3−pyridyl	2.7	
C_6		
$-C_6H_5$	1.4	0.8

Table 4.2 (*Contd.*)

	$-\Delta pK_a$	
Substituent	α-position‡	β-position§
C_7		
$-COC_6H_5$		2.2
C_n		
$-CH{=}CR_2$	~1.0	~0.5
$-C{\equiv}CR$	~2.0	~1.0
−NHR	~1.7	~0.9
$-NR_2$	~1.7	~0.9
−OR		1.2
−COR		1.6
−COOR	3.0	1.3
−OCOR		~1.7
−SR	~3.5	1.4

† Typical primary, secondary and tertiary amines have pK_a values of 10.77, 11.15, 10.5, respectively, at 20°. Ring formation increases pK_a by 0.2, but N-methylation decreases it by 0.2 relative to other N-alkylations.
‡ Substituent on a carbon atom next to the basic centre.
§ Substituent two carbon atoms away from the basic centre.
¶ Calculated directly from experimental pK_a values. Other values for the β-position calculated using Equation (4.2).
‖ Adjacent to a primary or secondary amine.
* Adjacent to a tertiary amine.

Table 4.3 *Base-weakening* ($-\Delta pK_a$) *effects of substituents commonly attached to β- and γ-positions on pyridines* (*Jaffe and Lloyd Jones*, 1964).

	$-\Delta pK_a$	
Substituent	β-	γ-
C_0		
−Br	2.33	
−Cl	2.33	1.45
−F	2.20	
−I	1.92	
−OH	0.43	
$-NH_2$	−0.86	−3.94
$-NO_2$	3.42	
$-SO_3^-$	1.85	1.67
C_1		
$-CH_3$	−0.51	−0.85
−CN	3.84	

Table 4.3 (*Contd.*)

	$-\Delta pK_a$	
Substituent	β-	γ-
$-COO^-$	0.52	0.39
$-OCH_3$	0.35	1.39
$-SCH_3$	0.78	−0.74
$-CONH_2$	1.72	
C_2		
$-C_2H_5$	−0.53	−0.85
$-COCH_3$	1.99	
$-NHCOCH_3$	0.77	0.64
C_3		
$-CH(CH_3)_2$	−0.55	−0.85
C_4		
$-C(CH_3)_3$	−0.65	−0.82
C_7		
$-NHCOC_6H_5$	1.43	0.09

4.2 Examples of the use of ΔpK_a values

4.2.1 Bis(2-chloroethyl)(2-methoxyethyl)amine

$$CH_3OCH_2CH_2N(CH_2CH_2Cl)_2$$

pK_a of tertiary amine	10.5 (Table 4.2, footnote §)
ΔpK_a for $-OCH_3$ on β-carbon	−1.2 (Table 4.2)
ΔpK_a for $-Cl$ on β-carbon	−2.1 (Table 4.2)
ΔpK_a for $-Cl$ on β-carbon	−2.1
Predicted $pK_a = 10.5 - 1.2 - 2.1 - 2.1$	= 5.10
Experimental pK_a	= 5.45

4.2.2 1-(4′-Hydroxycyclohexyl)-2-(isopropylamino)ethanol

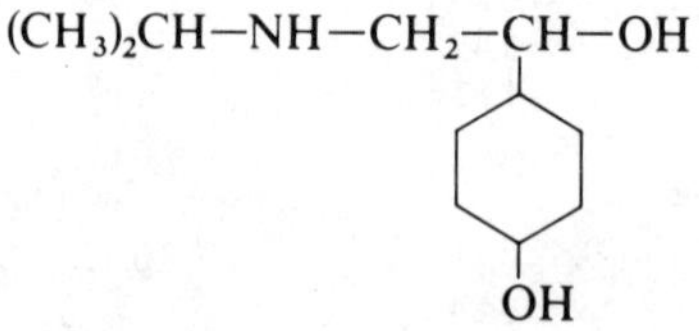

$$(CH_3)_2CH{-}NH{-}CH_2{-}CH{-}OH$$

OH

pK_a of secondary amine	11.15 (Table 4.2, footnote §)
ΔpK_a for $-OH$ on β-carbon	-1.1
ΔpK_a for $-OH$ on cyclohexyl, 4 carbons removed from the β-carbon $= -1.1 \times 0.4† \times 0.4 \times 0.4 \times 0.4 \times 0.4 \times 2‡$	$= -0.06$
Predicted $pK_a = 11.15 - 1.1 - 0.06$	$= 9.99$
Experimental pK_a	$= 10.23$

4.2.3 2-Aminocycloheptanol

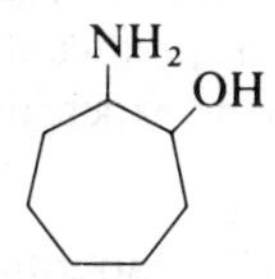

pK_a of primary amine	10.77 (Table 4.2, footnote §)
ΔpK_a for $-OH$ on β-carbon	-1.1
Predicted pK_a	$= 9.67$
Experimental pK_a	$= 9.25$

4.2.4 N,N-Dimethyl-2-butyn-1-amine

$$(CH_3)_2NCH_2C\equiv CCH_3$$

pK_a of tertiary amine	10.5
Decrease because of two N-methyls	-0.4 (Equation (3.4))
ΔpK_a for $R-C\equiv C-$ on α-carbon	~ -2.0
Predicted pK_a	~ 8.1
Experimental pK_a	8.28

4.2.5 5-Amino-3-azapentanol

$$H_2N-CH_2-CH_2-NH-CH_2-CH_2-OH$$

† Attenuation factor = 0.4 (Equation 4.3).
‡ Substituent effects operate along all chains. See Section 3.3.

(a) *Protonation on terminal nitrogen*

pK_a of primary amine	10.77
ΔpK_a for $-$NHR on β-carbon	-0.9
Predicted pK_a	= 9.87

(b) *Protonation on central nitrogen*

pK_a of secondary amine	11.15
ΔpK_a for $-$OH on β-carbon	-1.1
ΔpK_a for $-NH_2$ on β-carbon	-0.8
Predicted pK_a	= 9.25

Hence the stronger basic centre and the site of the first proton addition is $-NH_2$ (the experimental pK_a = 9.82).

With $-NH_2$ protonated, the predicted pK_a for the imino nitrogen = 11.15 − 3.6† − 1.1 = 6.45 (the experimental pK_a = 6.83).

4.2.6 5-Chloro-3-methyl-3-azapentanol

$$Cl-CH_2-CH_2-N(CH_3)-CH_2-CH_2OH$$

pK_a of tertiary amine	10.5
Decrease because of N-methyl	-0.2
ΔpK_a for $-$Cl on β-carbon	-2.1
ΔpK_a for $-$OH on β-carbon	-1.1
Predicted pK_a	7.1
Experimental pK_a	7.48

4.2.7 Hexamethylenetetramine

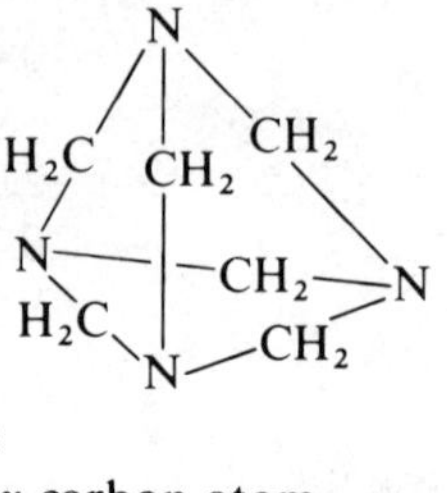

N
H_2C CH_2 CH_2
N $-CH_2-$ N
H_2C N CH_2

pK_a of tertiary amine	10.5
ΔpK_a for $-NR_2$ on α-carbon atom (3 times)	~ -5.1

† ΔpK_a for $-NH_3^+$ on β-carbon

Statistical factor (4 sites for proton addition	$+0.60(=\log 4)$
Predicted pK_a	$\sim$ 6.0
Experimental pK_a	6.2

4.2.8 2-Acetylbutanedioic acid

$$\begin{array}{c} HOOC-CH_2-\underset{\displaystyle COCH_3}{\underset{|}{CH}}-COOH \end{array}$$

The CH_3CO-group is acid-strengthening, so COOH-1 will be a stronger acid than COOH-4. The first proton therefore ionizes from COOH-1. The pK_a is predicted as follows:

pK_a of simple carboxylic acid	4.80
ΔpK_a for CH_3CO- on α-carbon	-1.10 (Table 4.1)
ΔpK_a for $-COOH$ on β-carbon $= 0.4(-1.37)$	$= -0.55$
Predicted pK_a	3.15
Experimental pK_a	2.86

For the second pK_a:

pK_a of simple carboxylic acid	4.80
ΔpK_a for CH_3CO- on β-carbon $= 0.4(-1.10)$	-0.44
ΔpK_a for $-COO^-$ on β-carbon $= 0.4(+0.61)$	0.24
Predicted pK_a	4.60
Experimental pK_a	4.57

The statistical effect is not applicable since the two acid groups are not equivalent.

4.2.9 Methionine

(a) *Prediction for zwitterionic form*:

$$CH_3-S-CH_2-CH_2-\underset{\displaystyle NH_3^+}{\underset{|}{CH}}-COO^-$$

pK_a of aliphatic carboxylic acid	4.80

ΔpK_a for $-NH_3^+$ on α-carbon	-2.43
ΔpK_a for $-SCH_3$ on γ-carbon = $-1.04(0.4)^2$	= -0.17
Predicted pK_a	= 2.20

pK_a of primary aliphatic amine	10.77
ΔpK_a for $-SCH_3$ on γ-carbon = $-1.6(0.4)$	= -0.64
ΔpK_a for $-COO^-$ on α-carbon	-0.8 (Table 4.2, footnote ‖)
Predicted pK_a	= 9.33

(b) *Prediction for non-zwitterionic form*:

$$CH_3-S-CH_2-CH_2-\underset{\displaystyle NH_2}{\underset{|}{CH}}-COOH$$

pK_a of aliphatic carboxylic acid	4.80
ΔpK_a for $-NH_2$ on α-carbon	-0.45
ΔpK_a for $-SCH_3$ on γ-carbon	-0.17 (See above)
Predicted pK_a	= 4.18

pK_a of primary aliphatic amine	10.77
ΔpK_a for $-COOH$ on α-carbon (not in Table 4.2; value for $-COOR$ used)	-3.0
ΔpK_a for $-SCH_3$ on γ-carbon = $-1.6(0.4)$	-0.64
Predicted pK_a	= 7.13

The experimental pK_a values for methionine are 2.17 and 9.20 which correspond closely with the values predicted for the zwitterionic form thus supporting the conclusion that methionine exists almost entirely in the zwitterionic form.

4.2.10 Piperazine

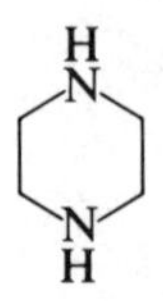

On the assumption† that substituent effects operate along all chains in a molecule, piperazine approximates to

$$R-NH-CH_2-CH_2-NH-CH_2-CH_2-NH-R$$

A correction† of 0.2 pK units must be added for ring closure, and, since in piperazine there are two identical sites for proton addition, a statistical factor ($\log 2 = 0.3$) must also be added.

pK_a of a secondary amine	=	11.15 (Table 4.2, footnote §)
ΔpK_a for $-NHR$ on β-carbon	=	−0.9 (Table 4.2)
Similarly, for the second $-NHR$, ΔpK_a	=	−0.9
ΔpK_a for ring closure	=	0.2
Statistical factor	=	0.3
Predicted pK_a	=	9.85
Experimental pK_a	=	9.82

4.3 The Taft equation

The additivity of free energy changes, as affecting the ionization constants of aliphatic and alicyclic species, is expressed more generally by the Taft equation:

$$pK = pK^0 - \rho^* \Sigma(\sigma^*) \tag{4.4}$$

where pK^0 is the ionization constant of the parent compound, ρ^* is a constant for the particular reaction and σ^* is a constant that is a characteristic of a given substituent. Table A.1 gives an extensive list of Taft σ^* constants, while Table A.2 includes a collection of published Taft equations. The standard procedure sets $\sigma^* = 0$ for the methyl group (giving $\sigma^* = 0.49$ for a hydrogen atom). The anomalies observed when hydrogen atoms on a basic nitrogen atom are replaced by methyl groups may be due to the decrease in hydrogen bonding to the solvent water. When σ^* for a substituent, $-R$, is known, σ^* for $-CH_2R$ can be estimated from the relation:

$$\sigma^* \text{ for } -CH_2R \sim 0.4 \times \sigma^* \text{ for } -R \tag{4.5}$$

The group $-\overset{|}{C}=\overset{|}{C}-$ is comparable with $-CH_2-$ in its attenuation of the effect of a substituent.

† See Section 3.3.

An estimate of σ^* for the group,

$$\begin{array}{c} X \\ | \\ -C-Y, \\ | \\ Z \end{array}$$

can be obtained from the σ^* values of $-CH_2X$, $-CH_2Y$ and $-CH_2Z$:

$$\sigma^* \text{ for } -CXYZ \sim \sigma^* \text{ for } -CH_2X + \sigma^* \text{ for } -CH_2Y + \sigma^* \text{ for } -CH_2Z. \quad (4.6)$$

The reliability of Relation (4.6) varies considerably as is shown in the Table on page 39.

The poor agreement for $-C(C_6H_5)_3$ may be related either to a massive steric effect around the methyl carbon or to a very large mesomeric effect of the three phenyls bonded to the central carbon. Conversely, with the small fluorine atoms, the high electron density on the carbon of $-CF_3$ may account for the observed discrepancy, which is greater for $-CF_3$ than for $-CCl_3$.

When, in Relation (4.6), X = Y = Z,

$$\sigma^*_{-CX_3} \sim 1/3 \times \sigma^*_{-CH_2X} \quad (4.7)$$

and similarly for $-CY_3$ and $-CZ_3$. Since $\sigma^*_{-CX_2Y} \sim \sigma^*_{-CH_2X} + \sigma^*_{-CH_2X} + \sigma^*_{-CH_2Y}$, it follows that

$$\sigma^*_{-CX_2Y} \sim 2/3(\sigma^*_{-CX_3}) + 1/3(\sigma^*_{-CY_3}) \quad (4.8)$$

In Taft equations, the reference compounds are those in which $-R$ is methyl for which σ^* is zero. Consider, as an example, the Taft equation for a protonated tertiary amine which ionizes thus:

$$R_1R_2R_3NH^+ \rightleftharpoons R_1R_2R_3N + H^+$$

From Table A.2,

$$pK_a = 9.61 - 3.30\,\Sigma(\sigma^*) \quad (4.9)$$

Here, the reference compound is $(CH_3)_3NH^+$. When methyls are replaced by other substituents, $\Sigma(\sigma^*)$ equals $\Sigma(\sigma^*$ for substituents other than methyl).

The Taft equation for protonated primary amines:

$$pK_a \text{ for } RNH_3^+ = 10.15 - 3.14(\sigma^*) \quad (4.10)$$

$-CH_2X$	σ^*	$-CH_2Y$	σ^*	$-CH_2Z$	σ^*	Sum	$-CXYZ$	σ^*	Error
$-CH_2H$	0	$-CH_2Cl$	0.94	$-CH_2Cl$	0.94	1.88	$-CHCl_2$	1.94	−0.06
$-CH_2CH_3$	−0.10	$-CH_2CH_3$	−0.10	$-CH_2CH_3$	−0.10	−0.30	$-C(CH_3)_3$	−0.30	0
$-CH_2H$	0	$-CH_2F$	1.10	$-CH_2F$	1.10	2.20	$-CHF_2$	2.05	+0.15
$-CH_2Cl$	0.94	$-CH_2Cl$	0.94	$-CH_2Cl$	0.94	2.82	$-CCl_3$	2.65	+0.17
$-CH_2C_6H_5$	0.27	$-CH_2C_6H_5$	0.27	$-CH_2C_6H_5$	0.27	0.81	$-C(C_6H_5)_3$	~0.4	+~0.4
$-CH_2Cl$	0.94	$-CH_2F$	1.10	$-CH_2F$	1.10	3.14	$-CClF_2$	2.59	+0.55
$-CH_2F$	1.10	$-CH_2F$	1.10	$-CH_2F$	1.10	3.30	$-CF_3$	2.61	+0.69

applies to acids with structure, RNH_3^+. When hydrogens are replaced by organic radicals, secondary and tertiary amines are formed and the appropriate Taft equations apply:

$$pK_a \text{ for } R_1R_2NH_2^+ = 10.59 - 3.23\,\Sigma(\sigma^*) \qquad (4.11)$$

and

$$pK_a \text{ for } R_1R_2R_3NH^+ = 9.61 - 3.30\,\Sigma(\sigma^*) \qquad (4.12)$$

Less obviously, when hydrogens are replaced by non-organic radicals, e.g. hydroxyl, an estimate of the pK_a of compounds such as $RNH(OH)$ and $R_1R_2N(OH)$ can be obtained using Equations (4.11) and (4.12), respectively. Thus:

$$\text{predicted } pK_a \text{ for } CH_3NH(OH) = 10.54 - 3.23(\sigma^* \text{ for } -CH_3 + \sigma^* \text{ for } -OH)$$
$$= 10.59 - 3.23(0 + 1.34) = 6.26$$

$$\text{predicted } pK_a \text{ for } (CH_3)_2N(OH) = 9.61 - 3.30(0 + 0 + 1.34) = 5.19$$

The experimental values are 5.96 and 5.20, respectively.

4.4 Examples of the use of Taft equations

4.4.1 2-(Methylamino)acetamide

$$CH_3-NH-CH_2-CONH_2$$

pK_a of $R_1R_2NH_2^+ = 10.59 - 3.23\,\Sigma(\sigma^*)$ (Table A.2)

$R_1 = -CH_3$, for which $\sigma^* = 0$ (Table A.1)

$R_2 = -CH_2CONH_2$, for which $\sigma^* = 0.4† \times \sigma^*$ for $-CONH_2$
$= 0.4 \times 1.68 = 0.67$

Predicted $pK_a = 10.59 - 3.23(0.67) = 8.43$
Experimental $pK_a = 8.31$

4.4.2 2-(Dimethylamino)ethyl acetate

$$CH_3-CO-OCH_2CH_2N(CH_3)_2$$

pK_a of $R_1R_2R_3NH^+ = 9.61 - 3.30\,\Sigma(\sigma^*)$ (Table A.2)

$R_1 = R_2 = -CH_3$ for which $\sigma^* = 0$

† From Equation 4.5.

$R_3 = -CH_2CH_2OCOCH_3$, for which $\sigma^* = 0.4 \times 0.4 \times \sigma^*$ for $-OCOCH_3$

$= 0.4 \times 0.4 \times 2.56$

~ 0.41

Predicted $pK_a = 9.61 - 3.30(0.41) \quad = 8.26$

Experimental $pK_a \quad = 8.35$

4.4.3 2,3-Dihydroxy-2-hydroxymethylpropanoic acid

$$\begin{array}{c} CH_2OH \\ | \\ HO-CH_2-C-COOH \\ | \\ OH \end{array}$$

pK_a for $R_1R_2R_3CCOOH = 5.10 - 0.81\,\Sigma(\sigma^*)$ (Table A.2)
σ^* for $-OH = 1.34$; $2(\sigma^*$ for $-CH_2OH) = 2(0.62)$
$= 1.24$ (Table A.1)

Predicted $pK_a = 5.10 - 0.81\ (2.58) = 3.01$
Experimental $pK_a = 3.29$

4.4.4 1,8-Diamino-3,6-dithiaoctane

$$H_2N-CH_2-CH_2-S-CH_2-CH_2-S-CH_2-CH_2-NH_2$$

In this molecule there are two equivalent basic groups and the statistical effect has to be taken into account after the Taft calculation.

$$pK_a \text{ of } RNH_3^+ = 10.15 - 3.14\,\sigma^* \qquad \text{(Table A.2)}$$

σ^* for the required R is not available in Table A.1. As an approximation, σ^* for $-CH_2CH_2SCH_2CH_3$ is given by:

$$0.4 \times 0.4 \times \sigma^* \text{ for } -SCH_2CH_3 = 0.4 \times 0.4 \times 1.56 = 0.25$$

Predicted pK_a for the numerically greater $pK_a = 10.15 - 3.14(0.25)$ + statistical factor $= 10.15 - 0.79 + 0.3 = 9.66$. Predicted pK_a for the numerically lesser $pK_a = 10.15 - 0.79 - 0.3 = 9.06$

Experimental values: 9.06 and 9.47

4.4.5 Chlorodifluoroacetic acid

$$ClF_2CCOOH$$

pK_a of RCOOH	$= 4.66 - 1.62\sigma^*$	(Table A.2)
σ^* for $-CClF_2$	$= 2.37$	(Table A.1)
Predicted pK_a	$= 4.66 - 1.62\ (2.37)$	
	$= 0.82$	
Experimental pK_a	$= 0.46$	

If σ^* had not been available in Table A.1, an estimate could be made using Relation (4.8):

$$\begin{aligned}\sigma^* \text{ for } -CClF_2 &\sim 1/3\ (\sigma^* \text{ for } -CCl_3) + 2/3\ (\sigma^* \text{ for } -CF_3)\\ &\sim 1/3\ (2.65) \qquad\qquad + 2/3\ (2.61)\\ &\sim 2.62\end{aligned}$$

4.4.6 Morpholine

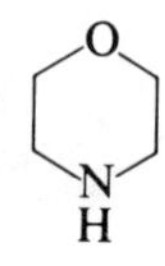

On the assumption‡ that substituent effects operate along all chains in a molecule, morpholine approximates to $CH_3O-CH_2-CH_2-NH-CH_2-CH_2-OCH_3$. A correction‡ of 0.2 p$K$ units for ring closure must be added to the value calculated from:

$$pK_a \text{ for } R_1R_2NH_2^+ = 10.59 - 3.23\,\Sigma\sigma^* \qquad \text{(Table A.2)}$$

$$\begin{aligned}\Sigma\sigma^* \text{ for } 2(CH_3OCH_2CH_2-) &= 2(0.4 \times 0.4 \times \sigma^* \text{ for } -OCH_3)\\ &= 2(0.4 \times 0.4 \times 1.81)\\ &= 0.58\end{aligned}$$

Predicted $pK_a = 10.59 - 3.23\ (0.58) + 0.2‡ = 8.92$

Experimental $pK_a = 8.36$

‡ See Section 3.3.

4.4.7 4-Morpholino-2,2-diphenylpentanenitrile

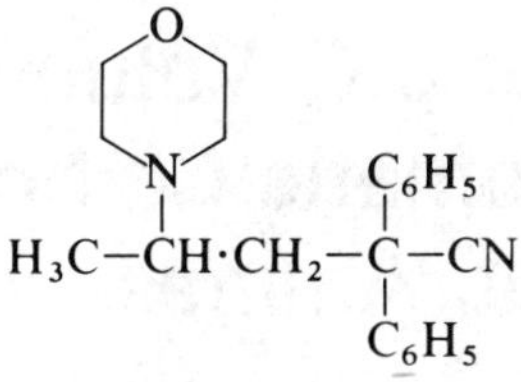

$$H_3C-\underset{\underset{\text{morpholino N}}{|}}{CH}\cdot CH_2-\underset{C_6H_5}{\overset{C_6H_5}{C}}-CN$$

pK_a of $R_1R_2R_3NH^+ = 9.61 - 3.30\,\Sigma(\sigma^*)$

(Table A.2)

pK_a of N-ethylmorpholine = 7.67

R_3 approximates to $-CH_2CH_2C(C_6H_5)_2(CN)$

Using Relation (4.6):

σ^* for $-C(C_6H_5)_2(CN) \sim 2(\sigma^*$ for $-CH_2C_6H_5)$
$+ (\sigma^*$ for $-CH_2CN)$
$\sim 2(0.27) + 1.30 \sim 1.84$

Then σ^* for R_3 $\sim 0.4 \times 0.4 \times 1.84 \sim 0.29$

σ^* for $-C_2H_5$ $= -0.10$

Difference in σ^* $= \sim 0.39$

Difference in pK_a $\sim 0.39 \times 3.30 = 1.29$

where 3.30 is the slope of the relation for a tertiary amine

Predicted pK_a $\sim 7.67 - 1.29$
~ 6.38

Experimental pK_a $= 6.05$

Chapter Five

Prediction of pK_a Values for Phenols, Aromatic Carboxylic Acids and Aromatic Amines

Within any class of substituted aromatic acid or base it is usually found that the acidic dissociation constant will vary in much the same way as that of the corresponding derivative in any other class. In most cases, these changes in acid strength, ΔpK_a, produced by a substituent in a *meta-* or *para-*position can be predicted by the Hammett equation.

5.1 The Hammett equation

Equilibrium and kinetic processes for many *meta-* and *para-* substituted aromatic compounds show a linear, additive dependence on free energy changes. Hammett (1940) found that if σ is a constant assigned to a particular substituent and ρ is a constant for a particular equilibrium, pK_a values of aromatic acids and bases can be expressed in the form

$$pK_a = pK_a^0 - \rho(\Sigma\sigma) \tag{5.1}$$

where pK_a^0 is the pK_a of the unsubstituted acid or base. For the ionization of substituted benzoic acids in water at 25° C:

$$pK_a = 4.20 - \Sigma\sigma \tag{5.2}$$

ρ being arbitrarily assigned the value, unity, for this system.

Table A.1 provides an extensive tabulation of Hammett σ and Taft σ^* constants, drawn mainly from the compilations of Hansch *et al.* (1973) and Exner (1978). Some of the Hammett equations that can be used for pK_a prediction are given in Tables A.2 and A.3. Most of the σ constants have been obtained from mono-substituted benzoic acids in water, so that the pK_a values of di-, tri-, or higher substituted acids can be predicted by summing the σ constants and subtracting the total from the pK_a of the parent acid.

A positive σ constant indicates acid-strengthening or base-weakening relative to hydrogen as a substituent (for which $\sigma = 0$). Conversely, a negative σ constant indicates acid-weakening or base-strengthening. If the reaction constant, ρ, for a particular equilibrium is known, the pK_a of any member of the series can be predicted by summing the σ constants for all of the substituents and inserting this value into the appropriate Hammett equation.

To improve the fit of Hammett-type relations, and to extend their range, the substituent parameter, σ, can be resolved into two components, σ_I (the polar or inductive effect) and σ_R (the mesomeric effect, the contributions due to 'resonance' or π-electron delocalization)†. By definition, σ_I is proportional to Taft's σ^*, the empirical relation being

$$\sigma_{meta} = 0.217\sigma^* - 0.106 \tag{5.3}$$

In the benzene series,

$$\sigma_R = \sigma_{para} - \sigma_I \tag{5.4}$$

The σ_R values can be further refined into σ^- values for electron-withdrawing substituents conjugated with electron-donating sidechains, and σ^+ values for electron-donating substituents conjugated to electron-withdrawing sidechains. On this basis the observed σ value is a composite of inductive and mesomeric effects, and will depend on the contribution of each as given by α and β in

$$\sigma_X = \alpha\sigma_{I,X} + \beta\sigma_{R,X} \tag{5.5}$$

Where mesomeric effects are inhibited, for example by interposing a $-CH_2CH_2-$ group between reaction centre and substituent, σ_X is essentially inductive in character. Dual parameter equations of this type have been widely used in recent years to evaluate substituent effects, using a blend of inductive and mesomeric contributions. They can cope with wide variations in resonance energies but it is difficult, *a priori*, to specify the values of α and β that are appropriate to a particular situation and hence to define values of σ for a particular reaction.

Yoshioka *et al.* (1962) have formulated the equation

$$\log K = \log K_0 + \rho[\sigma_{m,p} + r(\sigma_p^- - \sigma_p)] \tag{5.6}$$

where $\sigma_{m,p}$ are the ordinary Hammett constants and r is a variable

† To avoid confusion with σ_{meta}, the mesomeric effect is denoted by σ_R.

parameter. When $r = 0$ the expression reduces to the Hammett equation but when $r = 1$, σ_p^- replaces σ_p. When r lies between these limits, values are intermediate.

These σ_p^- values apply when there is enhanced mesomeric interaction and are used when the reaction centre is directly conjugated with the substituent. For substituents on anilines and phenols, σ_p^- constants are required. There are similar constants, σ_α^-, for substituted pyridines. These constants correspond to the 'special' σ constants listed in Table A.4. As an example, for the acetyl group σ_{para} is 0.50 for benzoic acids, σ_p^- is 0.84 for phenols and 0.81 for anilines, and σ_α^- is 0.28 for pyridines.

In these instances, σ_p values are not constant for different reactions, mainly because of variations in the extent to which the effect is transmitted from one end of the molecule to the other ('through-resonance'). The σ values for charged groups can also vary with the system. Thus σ_{para} for 4-$N(CH_3)_3^+$ is 1.8 in $ArCH_2CH_2COOH$, 1.2 in $ArCH_2COOH$ and 0.8 in $ArNH_3^+$.

We have assumed throughout that the ρ value for the reaction of a series of compounds having the substituents R′ and R″, where R′ is varied but R″ remains constant for the given series, is the same as the ρ value for a similar series of compounds having a different R″.

We have preferred to retain the simplicity of the original Hammett equation but to use these 'special' σ constants in those cases where σ_{para} values vary with the nature of the ionization reaction.

For a fuller discussion of the application of modified σ constants to acid–base equilibria, see Exner (1972).

5.2 Special σ_{ortho} and σ_{para} values of aromatic acids and bases

The influence of a substituent in an *ortho* position on an organic reaction varies considerably. Differences in proximity effects of a substituent on an acidic or basic centre include field effects operating strongly at short distances, steric effects making possible hindrance to coplanarity and inhibition of solvation, and possible hydrogen-bond formation; these contributions are additional to the usual inductive and resonance contributions.

Nevertheless, for any given reaction, such as proton addition or removal, for a specified class such as phenols, carboxylic acids or amines, the overall effect of an *ortho* substituent should not be greatly modified by other substituents and the apparent σ_{ortho} constants should be additive with σ_{meta} and σ_{para} constants. Table A.5 lists four sets of apparent σ_{ortho} constants for use with aromatic carboxylic acids, phenols, anilines and pyridines. It must be emphasized that these σ_{ortho} constants apply only for proton equilibria for the classes of compound indicated, but by their introduction the usefulness of the Hammett equation for predicting pK_a values is considerably extended.

Using the appropriate constants, pK_a values of 25 substituted benzoic acids were predicted on average to within ± 0.07 pH units (Clark and Perrin, 1964). Similarly, the pK_a values of 11 substituted anilines bearing two to four substituents and varying in pK_a value from 5.26 to -9.4 were predicted on average to within ± 0.28 (Clark and Perrin, 1964).

Because the pK_a values of substituted phenols and anilines fit Hammett equations using the same σ constants, an equation can be written relating the pK_a values of phenols to the pK_a values of anilines (Robinson, 1964):

$$pK^{subst}_{phenol} = 6.458 + 0.721\, pK^{subst}_{aniline} \tag{5.7}$$

This permits the prediction of the pK_a values of phenols from the pK_a values of anilines, and vice versa, and is useful in cases where *ortho* substituents are present because the steric interaction is less than with carboxylic acids. Thus, from the pK_a of 2,4-dichloroaniline ($= 2.05$), the pK_a of 2,4-dichlorophenol is predicted to be 7.94; the experimental value is 7.85.

5.3 Examples of the use of Hammett equations

5.3.1 4-Chloro-3,5-dimethylphenol

OH

H_3C CH_3

Cl

pK_a of a phenol $= 9.92 - 2.23\,\Sigma\sigma$ (Table A.2)

σ_{meta} for $-CH_3$ $= -0.06$; σ_{para} for $-Cl = 0.24$ (Table A.1)

Predicted pK_a $= 9.92 - 2.23\ (0.24 - 0.06 - 0.06)$
$= 9.70$
Experimental pK_a $= 9.71$

5.3.2 Pyrogallol

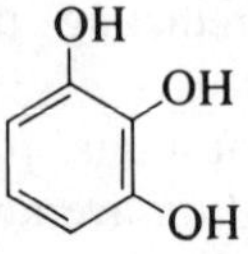

pK_a of a phenol $= 9.92 - 2.23\,\Sigma\sigma$ (Table A.2)
σ_{ortho} for $-OH$ $= 0.04$; σ_{meta} for $-OH = 0.13$;
σ_{meta} for $-O^-$ $= -0.47$ (Table A.1)

The hydroxyls on carbon-1 and carbon-3 are equivalent;

pK_{a1} for $-OH$ on carbon-1 (or -3) $= 9.92 - 2.23(0.04 + 0.13)$
$- \log 2 (= 0.3)$
$= 9.24$
Alternatively, for pK_{a1} on carbon-2 $= 9.92 - 2.23(0.04 + 0.04)$
$= 9.74$

Thus, the first ionization is from $-OH$ on carbon-1 (or -3) with predicted $pK_{a1} = 9.24$.

On electrostatic grounds, loss of protons from hydroxyls on carbon-1 and -2, or carbon-2 and -3, would be unlikely. Thus pK_{a1} can be assigned to carbon-1 or -3, and pK_{a2} to carbon-3 or -1. Then:

Predicted $pK_{a2} = 9.92 - 2.23\ (\sigma_{ortho}$ for $-OH$
$+ \sigma_{meta}$ for $-O^-) + \log 2\ (= 0.3)$
$= 9.92 - 2.23\ (0.04 - 0.47) + 0.3$
$= 11.18$

Experimental values: $pK_{a1} = 9.12$; $pK_{a2} = 11.19$

5.3.3 Benzenehexol

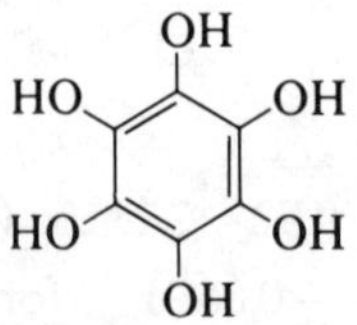

$$pK_a \text{ of a phenol} = 9.92 - 2.23\,\Sigma\sigma$$

A statistical factor of 0.78 (= log 6) has to be subtracted because there are six identical sites from which the proton could be removed.

σ_{ortho} for $-OH$ (phenols) = 0.04 (Table A.5)

σ_{meta} for $-OH$ = 0.13 (Table A.1)

σ_{para} for $-OH$ (phenols) = 0.03 (Table A.4)

Predicted pK_{a1} $= 9.92 - 2.23\,(2\sigma_{ortho} \text{ for } -OH + 2\sigma_{meta} \text{ for } -OH + \sigma_{para} \text{ for } -OH) - \log 6$

$= 9.92 - 2.23(0.08 + 0.26 + 0.03) - 0.78$

$= 8.31$

Experimental pK_{a1} $= 9.0$

5.3.4 Picric acid

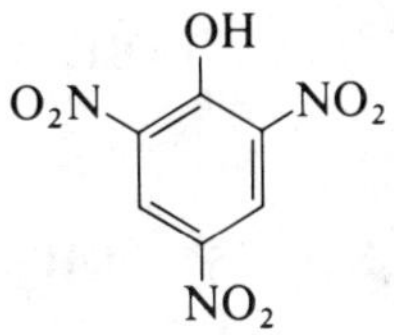

OH
O_2N NO_2
NO_2

pK_a of a phenol $= 9.92 - 2.23\,\Sigma\sigma$

σ_{ortho} for $-NO_2$ (phenols) = 1.40 (Table A.5)

σ_{para} for $-NO_2$ (phenols) = 1.24 (Table A.4)

Predicted pK_a $= 9.92 - 2.23\,(1.40 + 1.40 + 1.24)$

$= 0.91$

Experimental pK_a $= 0.33$

5.3.5 2,6-Dichloro-1,4-benzenediol

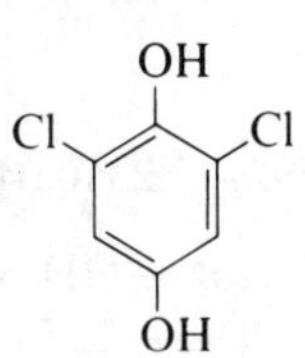

OH
Cl Cl
OH

pK_a of a phenol $= 9.92 - 2.23\,\Sigma\sigma$
σ_{ortho} for $-$Cl $= 0.68$; σ_{meta} for $-$Cl $= 0.37$;
σ_{para} for $-$OH $= 0.03$
σ_{para} for $-O^-$ $= -0.66$ (Table A.4)
Predicted pK_{a1} for
$-$OH on carbon-1 $= 9.92 - 2.23\,(0.68 + 0.68 + 0.03) = 6.82$

Alternatively, predicted pK_{a1} for $-$OH on carbon-4 $= 9.92 - 2.23\,(0.37 + 0.37 + 0.03) = 8.20$. Hence, the hydroxylic proton on carbon-1 is the stronger and will ionize first.

Predicted pK_{a1} $= 6.82$
Experimental pK_{a1} $= 7.30$
Predicted pK_{a2} (on carbon-4) $= 9.92 - 2.23\,(0.37 + 0.37 + -0.66)$
$= 9.74$
Experimental pK_{a2} $= 9.99$

5.3.6 3,4,5-Trihydroxybenzoic acid

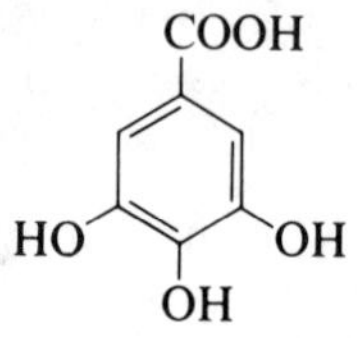

COOH
HO OH
OH

pK_a of an aromatic carboxylic acid $= 4.20 - \Sigma\sigma$ (Table A.2)
pK_a of a phenol $= 9.92 - 2.23\,\Sigma\sigma$ (Table A.2)
σ_{meta} for $-$OH $= 0.13$; σ_{para} for $-$OH $= -0.38$
σ_{ortho} for $-$OH (for phenol) $= 0.04$; σ_{para} for $-COO^-$ $= -0.05$
σ_{meta} for $-COO^-$ $= 0.09$

The first pK_a is for proton removal from the carboxyl group.

Predicted pK_{a1} $= 4.20 - (0.13 + 0.13 - 0.38) = 4.32$
Experimental pK_{a1} $= 4.46$
Predicted pK_{a2} for
$-$OH on carbon-4 $= 9.92 - 2.23\,(0.04 + 0.04 - 0.05) = 9.85$

Alternatively, predicted pK_{a2} for $-$OH on the equivalent positions, carbon-3 and carbon-5 $= 9.92 - 2.23\,(0.04 + 0.13 + 0.09) - \log 2$

(= 0.3) = 9.04. Hence, the hydroxylic proton on carbon-3 is the stronger acid and will ionize first.

$$\text{Predicted p}K_{a2} = 9.04$$
$$\text{Experimental p}K_{a2} = 8.78$$

5.3.7 4-Bromo-1,2-benzenedicarboxylic acid

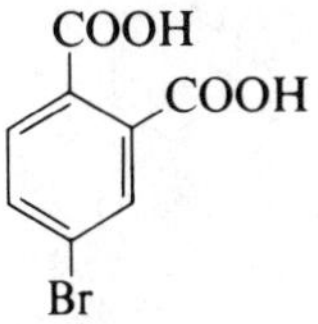

$\text{p}K_a$ for $-COOH$	$= 4.20 - \Sigma\sigma$	(Table A.2)
σ_{meta} for $-Br$	$= 0.39$; σ_{para} for $-Br$	$= 0.22$
σ_{ortho} for $-COOH$	$= 0.95$; σ_{ortho} for $-COO^-$	$= -0.91$

The greater positive value of σ_{meta} for $-Br$ compared with the σ_{para} value, means that the meta-Br has a greater acid-strengthening effect and the $-COOH$ on carbon-2 is the stronger acid (alternatively, calculate $\text{p}K_{a1}$ for the two possibilities). Hence:

$$\text{Predicted p}K_{a1} = 4.20 - (0.95 + 0.39) = 2.86$$
$$\text{Experimental p}K_{a1} = 2.50$$
$$\text{Predicted p}K_{a2} = 4.20 - (-0.91 + 0.22) = 4.89$$
$$\text{Experimental p}K_{a2} = 4.60$$

5.3.8 4-Hydroxy-3,5-dimethoxybenzoic acid

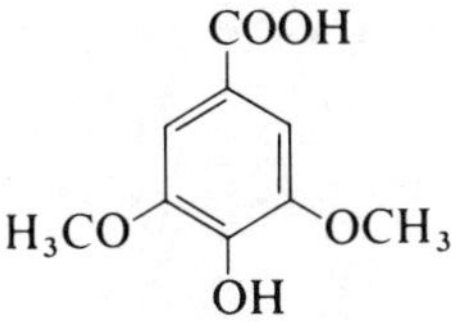

$\text{p}K_a$ for a benzoic acid	$= 4.20 - \Sigma\sigma$	(Table A.2)
σ_{meta} for $-OCH_3$	$= 0.11$; σ_{para} for $-OH$ $= -0.38$	

Predicted pK_{a1}	$= 4.20 - (0.11 + 0.11 - 0.38)$
	$= 4.36$
Experimental pK_{a1}	$= 4.34$

5.3.9 3-Iodo-4-methylthioaniline

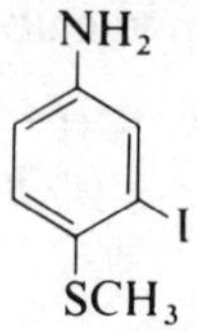

NH_2 / I / SCH_3

pK_a for anilines	$= 4.58 - 2.88\,\Sigma\sigma$	(Table A.2)
σ_{meta} for $-I$	$= 0.35$	(Table A.1)
σ_{para} for $-SCH_3$ (anilines)	$= 0.08$	(Table A.4)
Predicted pK_a	$= 4.58 - 2.88(0.35 + 0.08) = 3.34$	
Experimental pK_a	$= 3.44$	

5.3.10 4-Bromo-3-nitroaniline

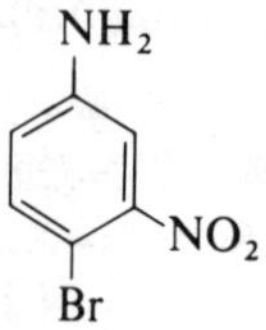

NH_2 / NO_2 / Br

pK_a for anilines	$= 4.58 - 2.88\,\Sigma\sigma$	(Table A.2)
σ_{para} for $-Br$	$= 0.22$; σ_{meta} for $-NO_2$	$= 0.74$
Predicted pK_a	$= 4.58 - 2.88(0.22 + 0.74)$	$= 1.82$
Experimental pK_a	$= 1.80.$	

Chapter Six

Further Applications of Hammett and Taft Equations

6.1 Prediction of pK_a values of heteroaromatic acids and bases

The success of the method of prediction based on the additivity of increments of pK_a values along chains in a molecule is not evidence of the correctness of the assumption that inductive effects are of overriding importance. Ring formation, by constraining the substituent in much closer proximity to the acidic or basic centre, leads to a stronger field effect than would be exerted by the same substituent attached at the free end of a chain, and may make a major contribution.

The pK_a values of *β*- and *γ*-substituted pyridines are roughly linear when plotted against σ constants for *m*- and *p*-substituted anilines (Jaffé and Doak, 1955). Slightly better plots are obtained if two lines are drawn, one being for substituents with $+M$ effects and the other for substituents with $-M$ effects. The inductive components (σ_I) for *m*-substituents are about the same for anilines, naphthylamines, pyridines and quinolines (Bryson, 1960*a*). Consequently, the pK_a values of substituted quinolines and *α*-naphthylamines can be interrelated by an equation

$$(\mathrm{p}K_a)^{\mathrm{subst}}_{\mathrm{quin}} = 2.1\,(\mathrm{p}K_a)^{\mathrm{subst}}_{\mathrm{naphth}} - 3.1 \tag{6.1}$$

usually to within ± 0.3 pH unit. Similarly for substituted isoquinolines and *β*-naphthylamines:

$$(\mathrm{p}K_a)^{\mathrm{subst}}_{\mathrm{isoquin}} = 2.1\,(\mathrm{p}K_a)^{\mathrm{subst}}_{\mathrm{naphth}} - 3.7 \tag{6.2}$$

so that knowledge of the pK_a of either of these classes of compounds provides an estimate of the pK_a of its analogue.

The key to predicting pK_a values of heterocyclic acids and bases is provided by the suggestion (Jaffé, 1953) that ring nitrogens in such molecules can be assigned Hammett constants so that the molecules are treated like substituted benzenes. From these constants, given in Table A.6, a ring nitrogen in a heteroaromatic molecule is roughly equivalent in terms of inductive and mesomeric effects to a nitro group attached at the same site on a benzene ring. This generalization fails for pyridazines and for other bases in which there is a ring nitrogen next to the basic nitrogen. Pyridazine, cinnoline, phthalazine and 3,4-benzocinnoline are much stronger bases than predicted. In bases that are polyaza heteroaromatics all but the most basic ring nitrogen atom can be replaced by a carbon bearing a nitro group. In acids that are polyaza heteroaromatics, all of the nitrogen ring atoms can be so considered.

Thus, substituents in the β- and γ-positions on pyridine have, in general, σ constants similar to those for *m*- and *p*- in the benzene series, with special constants being assigned in cases where 'through-resonance' is important. Examples where these anomalies are observed are the groups $-OH$, $-CHO$ and $-NO_2$ in a γ-position on pyridine. Likewise, special σ constants for α-substituents parallel usage in the aniline series.

Where an amino group is attached to a ring carbon atom in a heteroaromatic nitrogen compound, such as 3-aminopyridine (*6.1*), protonation occurs on the aromatic ring nitrogen but when an aminoalkyl group is the substituent, as in 3-aminomethylpyridine (*6.2*), the proton is added to the exocyclic nitrogen.

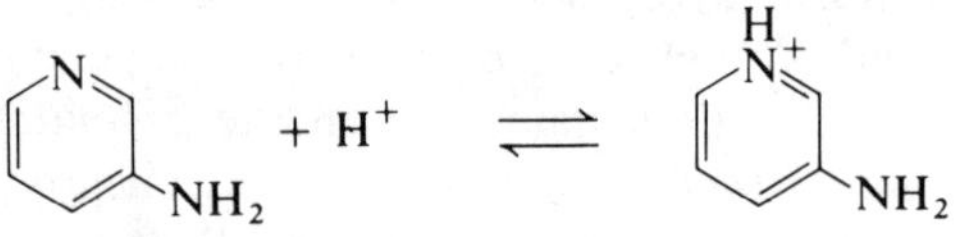

NH_2 + H^+ ⇌ H N+ NH_2

(*6.1*)

CH_2NH_2 + H^+ ⇌ N $CH_2NH_3^+$

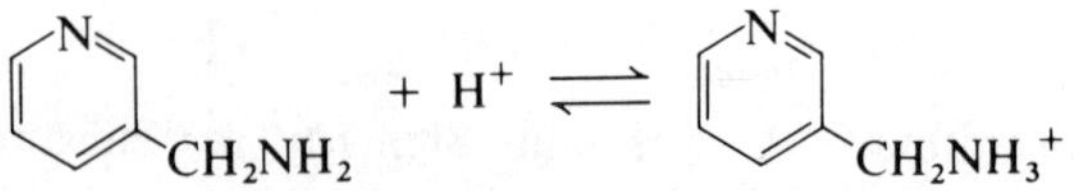

(*6.2*)

Although the present treatment correctly predicts the pK_a values of 3-hydroxypyridines it fails to reproduce the values for 2- and 4-hydroxypyridines. This is because the latter exist predominantly in the amide or pyridone form (6.3, 6.4) with proton ionization occurring from the nitrogen atom (*6.5, 6.6*). Similar structures exist for 2- and 4-

hydroxypyrimidines and other heterocycles where such tautomerism is possible, and also in the corresponding thiols. The amide form is much more weakly acidic, and hence more strongly basic, than the

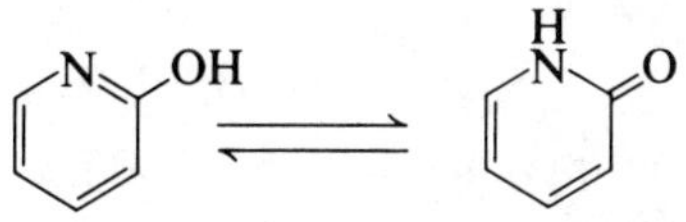

(6.3)

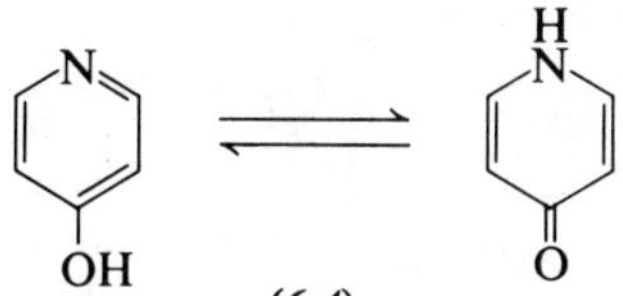

(6.4)

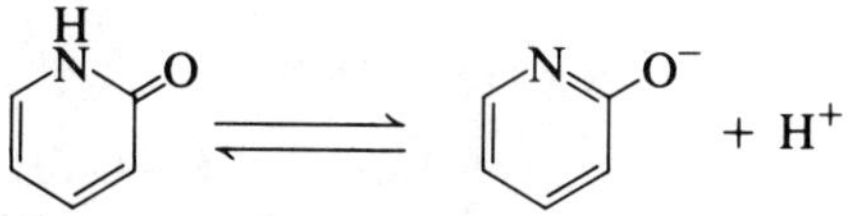

(6.5)

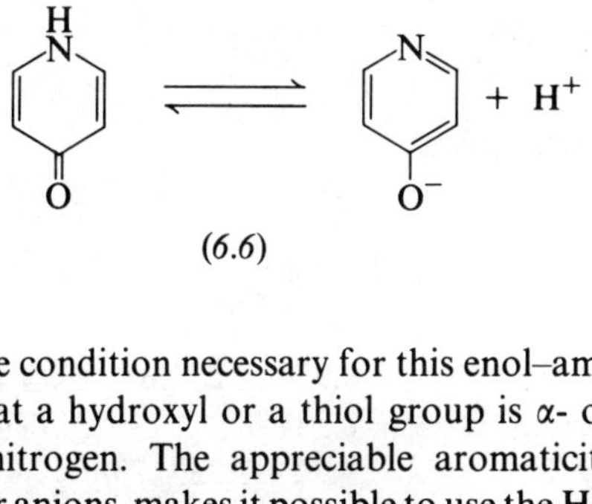

(6.6)

enol form. The condition necessary for this enol–amide tautomerism to occur is that a hydroxyl or a thiol group is α- or γ- to a doubly bound ring nitrogen. The appreciable aromaticity of pyridones, especially their anions, makes it possible to use the Hammett equation for predicting their pK_a values. For 1,2-dihydro-2-oxopyridines (2-pyridones) and 1,4-dihydro-4-oxopyridines (4-pyridones) the relevant equations are:

$$pK_a \text{ of 2-pyridones} = 11.65 - 4.28\,\Sigma\sigma \qquad (6.3)$$

$$pK_a \text{ of 4-pyridones} = 11.12 - 4.28\,\Sigma\sigma \qquad (6.4)$$

(The ρ values are the same as for ionization of azoles).

6.2 Worked examples: heteroaromatic acids and bases

6.2.1 3-Acetylpyridine

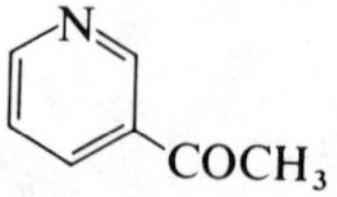

pK_a of a substituted pyridine $= 5.25 - 5.90\,\Sigma\sigma$ (Table A.3)
σ_{meta} for $-COCH_3$ $= 0.36$
Predicted pK_a $= 5.25 - 5.90\,(0.36)$
$= 3.13$
Experimental pK_a $= 3.18$

6.2.2 3-Bromo-5-methoxypyridine

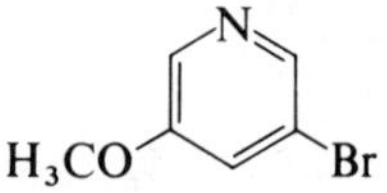

σ_{meta} for $-Br$ $= 0.39$; σ_{meta} for $-OCH_3 = 0.11$
Predicted pK_a $= 5.25 - 5.90\,(0.39 + 0.11)$ (See 6.2.1)
$= 2.30$
Experimental pK_a $= 2.60$

6.2.3 3-Hydroxypyridine

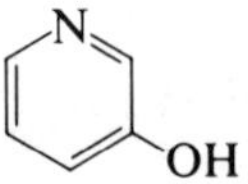

pK_a of a phenol $= 9.92 - 2.23\,\Sigma\sigma$ (Table A.2)
σ_{meta} for a ring N† $= 0.73$ (Table A.6)
Predicted pK_a $= 9.92 - 2.23(0.73)$
$= 8.29$
Experimental pK_a $= 8.74$
pK_a of a pyridine $= 5.25 - 5.90\,\Sigma\sigma$
σ_{meta} for $-OH$ $= 0.13$
Predicted pK_a $= 5.25 - 5.90(0.13)$
$= 4.48$
Experimental pK_a $= 4.80$

† Since pK_a of a pyridine ~ 5.25, the N is not protonated when considering the pK_a of a phenol.

6.2.4 Pyrimidine-2-carboxylic acid

N COOH
N

pK_a of a substituted benzoic acid	$= 4.20 - \Sigma\sigma$	(Table A.2)
σ_{ortho} for *each* ring N‡	$= 0.56$	
Predicted pK_a ($-COOH$)	$= 4.20 - (2 \times 0.56)$	
	$= 3.08$	
Experimental pK_a	$= 2.85$	
pK_a of a substituted pyridine	$= 5.25 - 5.90\,\Sigma\sigma$	
σ_{ortho} for $-COOH$	$= 0.51$	
σ_{meta} for a ring N	$= 0.73$	(Table A.6)
Statistical factor	$= +0.3$ (because of the two basic centres)	
Predicted pK_a	$= 5.25 - 5.90(0.51 + 0.73) + 0.3$	
	$= -1.8$	
Experimental pK_a	$= -1.1$	

6.2.5 4-Aminopyridazine

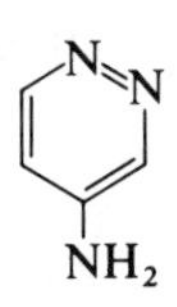

The first protonation is on a ring nitrogen (See Section 6.1).

$$pK_a \text{ of a substituted pyridine} = 5.25 - 5.90\,\Sigma\sigma$$

(a) *For N − 1 protonated*

σ_{ortho} for a ring N $= 0.56$ (Table A.6)

σ_{para} for $-NH_2$ $= -0.57$ (Table A.1)

Predicted pK_a $= 5.25 - 5.90(0.56 - 0.57)$

$= 5.31$

(b) *For N − 2 protonated*

σ_{ortho} for a ring N $= 0.56$

‡ The effect of a ring nitrogen is similar to a nitro group attached at the same site on a benzene ring; each nitrogen is consequently weakly basic.

$$\sigma_{meta} \text{ for } -NH_2 = 0.00$$
$$\text{Predicted p}K_a = 5.25 - 5.90(0.56) = 1.95$$

Therefore, the more basic ring N is N – 1 with

$$\text{Predicted p}K_a = 5.31$$
$$\text{Experimental p}K_a = 6.65$$

6.2.6 4-Amino-6-chloropyrimidine

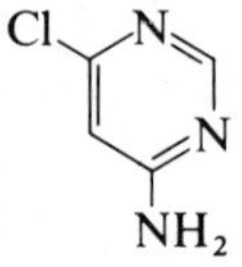

The first protonation is on a ring nitrogen.

$\text{p}K_a$ for a substituted pyrimidine $= 1.23 - 5.90\,\Sigma\sigma$ (Table A.3)

(a) *For protonation on N – 1*

σ_{para} for $-NH_2$ $= -0.57$ (Table A.1, note §)
σ_{ortho} for $-Cl$ $= 0.79$ (Table A.5, pyridines)
$\Sigma\sigma = 0.22$

(b) *For protonation on N – 3*

σ_{ortho} for $-NH_2$ $= -0.27$ (Table A.5)
σ_{para} for $-Cl$ $= 0.24$ (Table A.1)
$\Sigma\sigma = -0.03$

Since $\Sigma\sigma$ for protonation on N-3 is lower, N-3 is a stronger base with

$$\text{Predicted p}K_a = 1.23 - 5.90(-0.03) = 1.41$$
$$\text{Experimental p}K_a = 2.10$$

Alternatively, the calculation could be made as in Example 6.2.4, or by using the Hammett equation for a 4-amino-6-substituted pyrimidine given in Table 11.3.

6.2.7 1,2-Dihydro-6-chloro-4-methylpyridine-2(*1H*)-one

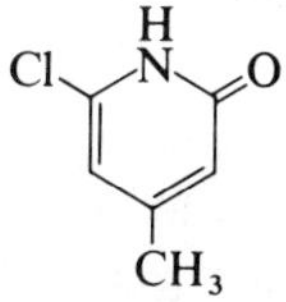

pK_a of a 2-pyridone	$= 11.65 - 4.28\,\Sigma\sigma$	(Equation 6.3)
σ_{ortho} for $-Cl$	$= 0.79$	
σ_{para} for $-CH_3$	$= -0.14$	
Predicted pK_a	$= 11.65 - 4.28(0.79 - 0.14)$	
	$= 8.9$	
Experimental pK_a	$= 8.06$	

Similarly, for 1,2-dihydro-5-iodopyridine-2(1*H*)-one:

Predicted pK_a	$= 10.15$
Experimental pK_a	$= 9.93$

For 1,2-dihydro-6-methoxypyridine-2(1*H*)-one

Predicted pK_a	$= 10.19$
Experimental pK_a	$= 9.47$

For 1,4-dihydro-3,5-dinitropyridine-4(1*H*)-one

Predicted pK_a	$= 4.79$	(using Equation 6.3)
Experimental pK_a	$= 4.56$	

6.2.8 Prediction of tautomeric ratio [2-pyridone]/[2-hydroxypyridine]

pK_a of 2-pyridone $= 11.65$ (Equation 6.3)

pK_a of 2-hydroxypyridine is calculated as follows:

pK_a of a phenol	$= 9.92 - 2.23\,\Sigma\sigma$	
σ_{ortho} for a ring N	$= 0.56$	(Table A.6)
Predicted pK_a	$= 9.92 - 2.23(0.56)$	
	$= 8.67$	

Therefore, predicted tautomeric ratio

[2-pyridone]/[2-hydroxypyridine]	$= 10^{11.65}/10^{8.67}$	(Equation 2.6)
	$= 955$	
Experimental value	$= 340$	

6.3 Prediction of pK_a values of alcohols, aldehydes and other weak acids

6.3.1 Alcohols

Insofar as alcohols can be regarded as being derived from carboxylic acids by replacement of an oxygen atom by two hydrogen atoms, alcohols and acids would be expected to show similar effects of substituents on pK_a values. For the ionization of alcohols:

$$RCH_2OH \rightleftharpoons RCH_2O^- + H^+$$

the experimental Taft equation at 25° is

$$pK_a = 15.9 - 1.42\,\Sigma\sigma^* \tag{6.5}$$

which fitted pK_a values to within about ± 0.2 pH unit for seven alcohols of the type, RCH_2OH (Ballinger and Long, 1960). (Equation (6.5) gives the pK_a of ethanol as 15.9.)

6.3.2 Hydrated aldehydes

Hydrated aldehydes, $RCH(OH)_2$, derived from alcohols by replacement of an $-H$ by an $-OH$, ionize thus:

$$RCH(OH)_2 \rightleftharpoons RCH(OH)O^- + H^+$$

The difference in σ^* constants for $-H$ and $-OH$ $(1.34 - 0.49)$, and allowance for a statistical factor of 0.3 (because there are now two equivalent acidic centres) leads to the Taft equation for the pK_a of a hydrated aldehyde:

$$pK_a = 14.4 - 1.42\,\Sigma\sigma \tag{6.6}$$

6.3.3 Thioalcohols

The pK_a values of thioalcohols at 25° are fitted by the Taft equations:

$$pK_a = 10.22 - 3.50\,\Sigma\sigma^*, \tag{6.7}$$

for $RSH \rightleftharpoons RS^- + H^+$, or

$$pK_a = 10.54 - 1.47\,\Sigma\sigma^* \tag{6.8}$$

for $RCH_2SH \rightleftharpoons RCH_2S^- + H^+$.

6.3.4 Substituted azoles (as acids)

Pyrroles and their aza- and annelated derivatives are very weak acids ionizing thus

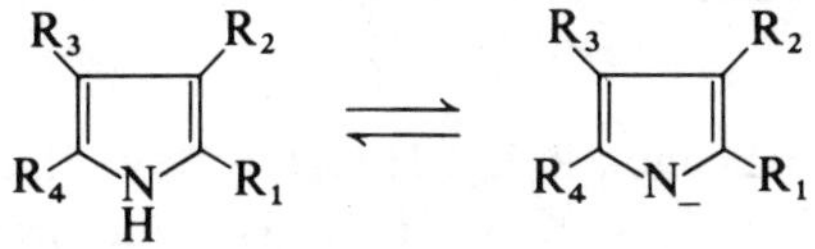

$$\text{(pyrrole, } R_1\text{–}R_4\text{, N–H)} \rightleftharpoons \text{(pyrrolide anion, } R_1\text{–}R_4\text{, N}^-\text{)} + H^+$$

for which

$$pK_a \sim 17.0 - 4.28\,\Sigma\sigma$$

where the Hammett σ values for ring $=N-$ groups are $\sigma_{ortho} = 0.56$ and $\sigma_{meta} = 0.73$. For 36 derivatives ranging from pyrazole to 6-methylthiopyrazole [5′,4′:4,5] pyrimidine, the average deviation from prediction was ± 0.4 pH unit (Barlin and Perrin, 1966).

6.4 Worked examples: alcohols, aldehydes and other weak acids

6.4.1 Glucose

Treated as a hydrated aldehyde

$$HOCH_2-CH(OH)-CH(OH)-CH(OH)-CH(OH)-CH(OH)_2$$

pK_a of a hydrated aldehyde	$= 14.4 - 1.42\,\Sigma\sigma^*$ (Equation 6.6)
σ^* for $-OH$ on carbon-2	$= 0.4\dagger \times (\sigma^*$ for $-OH)\ddagger$
	$= 0.4 \times 1.34 = 0.54$
σ^* for $-OH$ on carbon-3	$= 0.21$
σ^* for $-OH$ on carbon-4	$= 0.09$
σ^* for $-OH$ on carbon-5	$= 0.03$
σ^* for $-OH$ on carbon-6	$= 0.01$
$\Sigma\sigma^*$	$= 0.88$
Predicted pK_a	$= 14.4 - 1.42(0.88)$
	$= 13.1$
Experimental pK_a	$= 12.3$

† Equation 4.3; ‡ Table A.1

Alternatively, glucose treated as an alcohol:

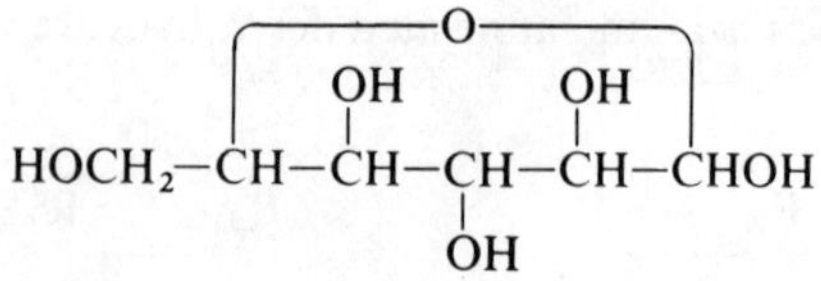

```
     ┌─────────────O─────────────┐
     │      OH          OH       │
     │      |           |        │
HOCH2—CH—CH—CH—CH—CHOH
                |
                OH
```

pK_a of an alcohol, $R_1R_2CH(OH) = 15.9 - 1.42\,\Sigma\sigma^*$
$-O-CH(CH_2OH)CHOHCHOH-$ on carbon-1
approximates to $-OCH(CH_3)CH_2CH_3$ for
which σ^* is available (= 1.62)
σ^* for $-OH$ on carbon-2 = 0.54
(See first alternative)
σ^* for $-OH$ on carbon-3 = 0.21
σ^* for $-OH$ carbon-4 = 0.09

Substituent on carbon-5 approximates to $-OCH_3$ separated by four carbons, so that $\sigma^* = (0.4)^4 \times 1.81 = 0.05$

σ^* for $-OH$ on carbon-6 = 0.01
$\Sigma\sigma^* = 2.52$
Predicted $pK_a = 15.9 - 1.42(2.52) = 12.3$
Experimental $pK_a = 12.3$

6.4.2 Fructose

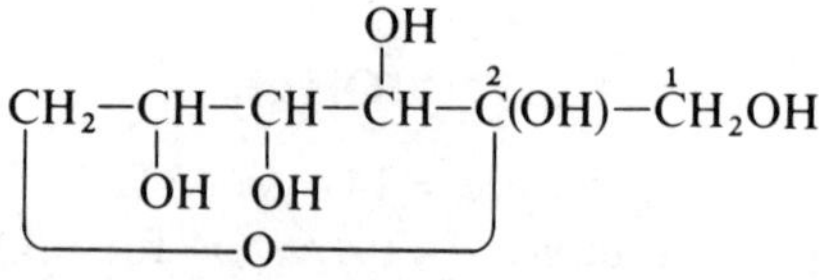

```
                   OH
                   |      2         1
CH2—CH—CH—CH—C(OH)—CH2OH
 |   |   |          |
 |   OH  OH         |
 └────────O─────────┘
```

Problem: Does the proton come from $-OH$ on carbon-1 or carbon-2?

pK_a of an alcohol, $RCH_2OH = 15.9 - 1.42\,\Sigma\sigma^*$

For ionization of the proton from the hydroxyl on carbon-1:

$-OCH_2CHOHCHOHCHOH-$ on carbon-2 approximates to
$-O(CH_2)_3CH_3$ for which $\sigma^* = 0.4 \times 1.68$ = 0.67
σ^* for $-OH$ on carbon-2 = 0.4×1.34 = 0.54
σ^* for $-OH$ on carbon-3 = 0.21

σ^* for $-OH$ on carbon-4 $= 0.09$
σ^* for $-OH$ on carbon-5 $= 0.03$

Substituent on carbon-6 approximates to $-OCH_3$ separated by five carbons, so that

$$\sigma^* = (0.4)^5 \times 1.81 = 0.02$$
$$\Sigma\sigma^* = 1.56$$

For ionization of the proton from the hydroxyl on carbon-2:

σ^* for $-OH$ on carbon-1 $= 0.4 \times 1.34 = 0.54$
σ^* for $-O(CH_2)_3CH_3$ on carbon-2 $= 1.68$
σ^* for $-OH$ on carbon-3 $= 0.54$
σ^* for $-OH$ on carbon-4 $= 0.21$
σ^* for $-OH$ on carbon-5 $= 0.09$

Substituent on carbon-6 approximates to $-OH$ for which

σ^* $= 0.03$
$\Sigma\sigma^*$ $= 3.09$

If the acidic centre is located on carbon-1, the predicted $pK_a = 13.7$ (using Equation 6.5) whereas if it is on carbon-2, the predicted $pK_a = 11.5$. The latter is the stronger acid and hence ionization from carbon-2 is the preferred reaction. The experimental pK_a at 25° = 12.15.

6.4.3 Comparison of the acidities of the alcoholic–OH groups of tartaric acid, lactic acid, malic acid and citric acid

pK_a for an alcohol, $R_1R_2CH(OH) = 15.9 - 1.42\,\Sigma\sigma^*$

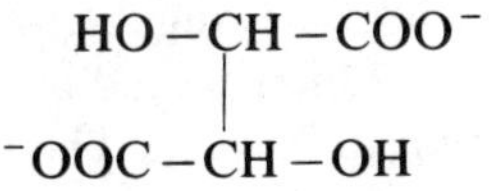

$HO-CH-COO^-$ / $^-OOC-CH-OH$ (the two CH carbons bonded to each other)

Tartrate ion

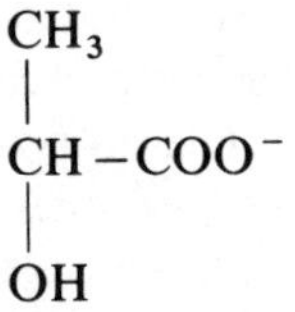

$CH_3-CH(OH)-COO^-$

Lactate ion

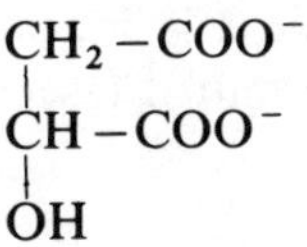

CH_2-COO^- / $CH-COO^-$ / OH (on the CH carbon)

Malate ion

CH_2-COO^- / $^-OOC-C-OH$ / CH_2-COO^- (both CH_2 groups on the central C)

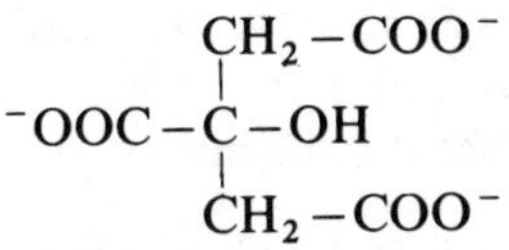

Citrate ion

	Tartrate	Lactate	Malate	Citrate
σ^* for $-COO^-$ on α-carbon	-1.06	-1.06	-1.06	-1.06
σ^* for $-COO^-$ on β-carbon	-0.42	—	-0.42	-0.42
σ^* for $-COO^-$ on β-carbon	—	—	—	-0.42
σ^* for $-OH$ on β-carbon	0.54	—	—	—
$\Sigma\sigma^*$	-0.94	-1.06	-1.48	-1.90
Predicted pK_a	16.9†	17.4	18.0	18.6

6.4.4 4-Phenylimidazole

HN N

pK_a for an aza derivative of a pyrrole $= 17.0 - 4.28\,\Sigma\sigma^*$ (Table A.2)

σ_{meta} for $=N-$ $= 0.73$
σ_{meta} for $-C_6H_5$ $= 0.05$
Predicted pK_a $= 17.0 - 4.28\ (0.73 + 0.05)$
$= 13.7$
Experimental pK_a $= 13.2$

6.5 Prediction of pK_a values of weak bases

Acetophenones substituted in the benzene ring, benzaldehydes and benzoic acids have very weak basic properties, and in sulphuric acid solutions they undergo protonation on oxygen. The pK_a values of substituted acetophenones (Steward and Yates, 1958) are given (usually to within ± 0.2 pH unit) by the equation

$$pK_a = -6.0 - 2.6\,\sigma \tag{6.10}$$

(the value of ρ is close to the expected value, 2.4‡, for protonation on the carbonyl oxygen if the effect is transmitted essentially quantitatively across the double bond).

† Includes a statistical factor (-0.3) for two $-OH$ groups.
‡ See page 87.

Similarly, the pK_a values of substituted benzaldehydes (Yates and Stewart, 1959) are reasonably well fitted by the equation

$$pK_a = -6.7 - 2.6\sigma \tag{6.11}$$

Protonation of benzoic acid appears to take place on the hydroxyl oxygen, so that the observed value of ρ is close to unity: the strengths of benzoic acids (Stewart and Granger, 1961) as bases, are well represented by the equation

$$pK_a = -7.26 - 1.2\sigma \tag{6.12}$$

Protonation of benzamides (Edward *et al*., 1960) shows them to be weak bases, the regression line for pK_a prediction being

$$pK_a = -2.15 - 1.3\sigma \tag{6.13}$$

This fitted the experimental data for 11 compounds to within ± 0.1 pH unit.

Chapter Seven

Some More Difficult Cases

7.1 Annelated rings as substituents

Polycyclic molecules containing only one acidic or basic centre can be treated as monocyclic systems with the annelated rings as substituents. The latter may be paraffinic (or heteroparaffinic), aromatic, π-deficient or π-excessive heteroaromatic.

The paraffinic rings can be replaced by alkyl groups attached to the ring containing the acidic or basic centre. Thus the pK_a (= 10.17) of bornylamine (*7.1*) is approximated, for example, by the pK_a values (*cis* = 10.35, *trans* = 10.48) of 1-amino-2-isopropyl-5-methylcyclohexane (*7.2*) and even that of ethylamine (= 10.70)

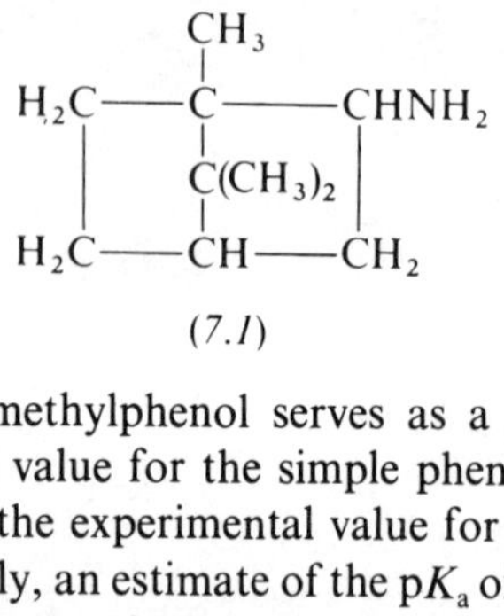

(*7.1*) 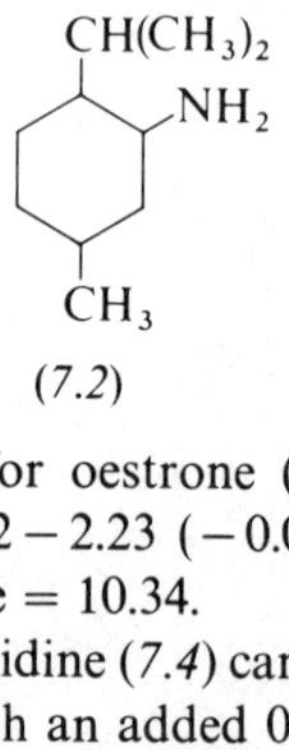 (*7.2*)

3,4-Dimethylphenol serves as a model for oestrone (*7.3*). The predicted value for the simple phenol = 9.92 − 2.23 (−0.06 − 0.14) = 10.37; the experimental value for oestrone = 10.34.

Similarly, an estimate of the pK_a of quinuclidine (*7.4*) can be made from the pK_a of triethylamine (= 10.75) with an added 0.3 for the formation of two rings through the nitrogen as described in Section 3.2.1. Predicted value = 11.05; experimental value = 10.80.

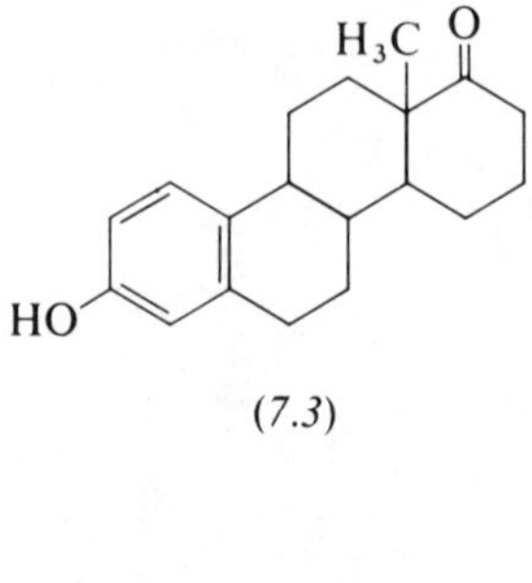

(*7.3*) 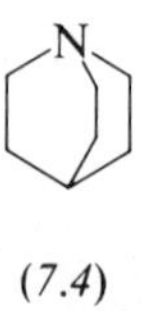 (*7.4*) 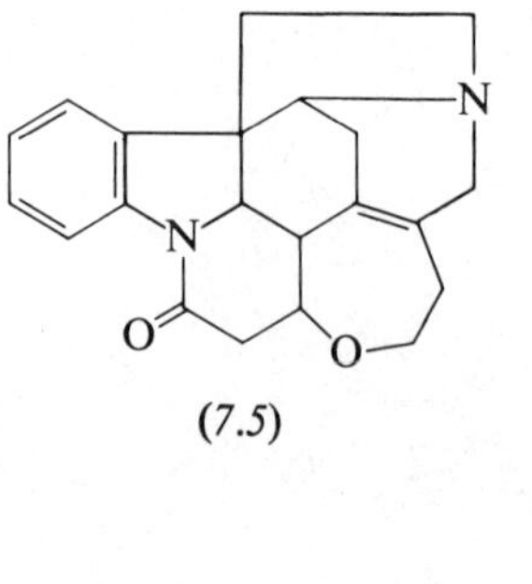 (*7.5*)

When the rings around the nitrogen atom are strained the experimental pK_a is less than predicted. For example, strychnine (*7.5*) has $pK_a = 8.26$, but the predicted value based on triethylamine as a model with two rings joined through the basic nitrogen is 11.05 (the second nitrogen is part of an amide and hence is only very weakly basic).

A 2-, 3- or 4-substituted α-naphthylamine can be looked upon as the corresponding aniline to which a substituent (a benzene ring) has been fused across positions 5- and 6. Similarly, a 1-, 3- or 4-substituted β-naphthylamine derives from aniline by fusion of a benzene ring across positions 4- and 5. Provided that the annelated rings do not, themselves, bear substituents they can thus be assigned Hammett σ constants and treated as substituents. Such Hammett constants are special to the total ring system under consideration. Thus for annelated benzene as a substituent of aniline in α-naphthylamine, $\sigma = +0.24$, whereas, as a substituent of pyridine in quinoline, $\sigma = +0.06$. Further Hammett constants for fused rings are given in Table 7.1. As an example, the pK_a of 3-hydrazinoquinoline (= 4.85) can be approximated using the Hammett equation for

Table 7.1 *Hammett sigma constants for fused rings*

$\Sigma\sigma$ for substituted	=	$\Sigma\sigma$ for substituted	*plus*	σ for the unsubstituted fused ring
α-naphthylamine	=	aniline	+	0.24
β-naphthylamine	=	aniline	+	0.08
α-naphthoic acid	=	benzoic acid	+	0.50
β-naphthoic acid	=	benzoic acid	+	0.04
α-naphthol	=	phenol	+	0.28
β-naphthol	=	phenol	+	0.11
quinoline	=	pyridine	+	0.06
isoquinoline	=	pyridine	−	0.02
acridine (*A*)	=	pyridine	−	0.04
benzo[f]quinoline (*B*)	=	pyridine	+	0.02
benzo[g]quinoline (*C*)	=	pyridine	+	0.04
benzo[h]quinoline (*D*)	=	pyridine	+	0.16
phenanthridine (*E*)	=	pyridine	+	0.12
1-aminoanthracene (*F*)	=	aniline	+	0.17
1-aminofluorene (*G*)	=	aniline	+	0.24
2-aminofluorene (*H*)	=	aniline	−	0.01
3-aminofluorene (*I*)	=	aniline	−	0.25
4-aminofluorene (*J*)	=	aniline	+	0.41
benz[a]acridine (*K*)	=	pyridine	+	0.09

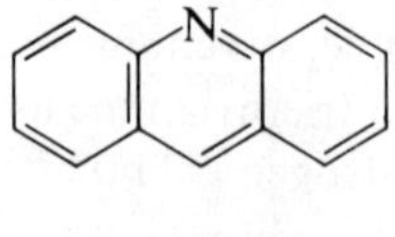

pyridines and inserting the σ constant for the 3-hydrazino group and the appropriate σ for annelated benzene:

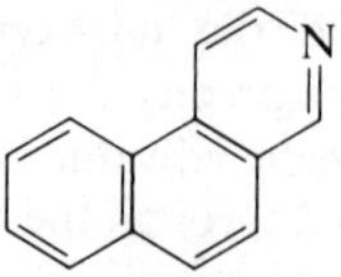

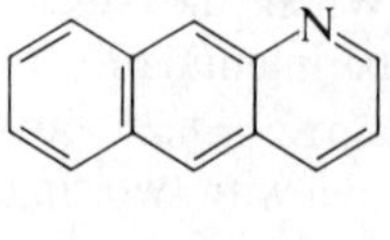

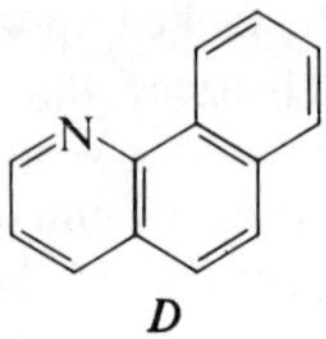

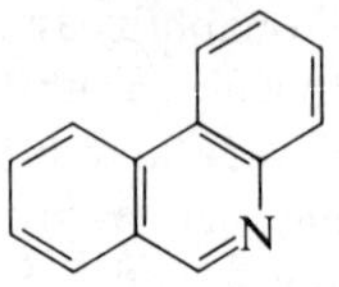

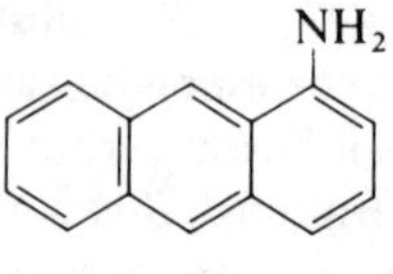

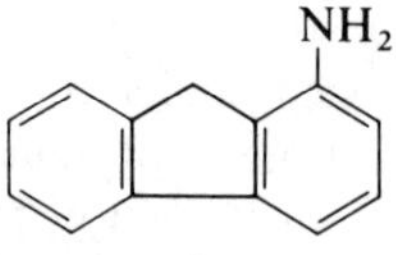

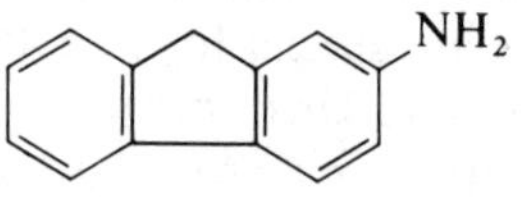

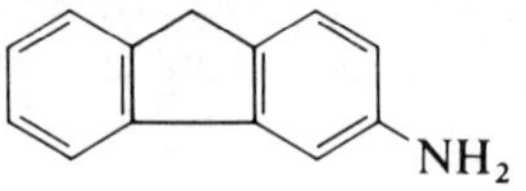

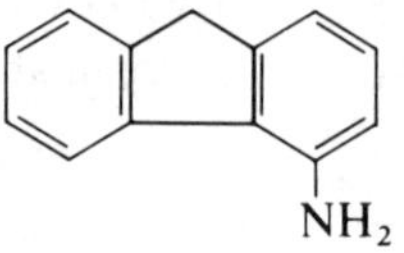

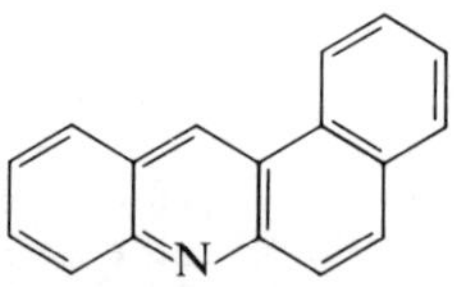

$$\text{predicted p}K_a = 5.25 - 5.90\,(-0.02 + 0.06) = 5.01.$$

In a comparison of the predicted pK_a values from substituted pyridines for 17 quinolines and 10 isoquinolines, the average fit was ± 0.32 pK units (Clark and Perrin, 1964). Similarly, the pK_a values of 13 substituted α-naphthylamines ranging from 0.54 to 5.87 were predicted within ± 0.13 (Clark and Perrin, 1964). The pK_a values predicted for substituted β-naphthylamines are somewhat lower for substituents with $-$R effects and somewhat higher for substituents with $+R$ effects: the deviation is usually less than $\pm 0.2\,\mathrm{p}K$ unit (Bryson, 1960). The $\mathrm{p}K_a$ values of a set of 17 1- and 2-naphthols were predicted on average within ± 0.08 from the corresponding phenols (Barlin and Perrin, 1966).

For systems containing more aromatic rings, similar additivities are assumed. Thus σ for the rings in benzo[c]acridine comprises σ ($= 0.06$) for one benzene ring fused to a pyridine (as in quinoline) and σ ($= 0.16$) for a naphthalene ring fused at its 1:2-position to positions 2,3 of pyridine (as in benzo[h]quinoline). The $\mathrm{p}K_a$ of benzo[a]-phenanthridine (*7.6*) is approximated using the Hammett

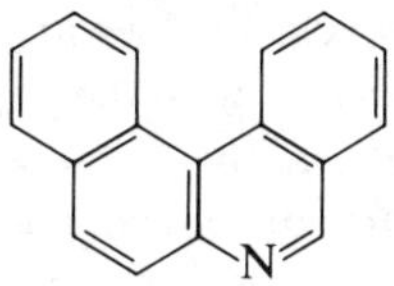

N

(*7.6*)

equation for a substituted pyridine and inserting σ constants for naphthalene fused to pyridine as in benzo[f]quinoline ($+0.02$) and benzene fused to pyridine as in isoquinoline (-0.02):

$$\text{predicted p}K_a = 5.25 - 5.90\ (0.02 - 0.02) = 5.25$$

The experimental value is 4.48.

It is a useful approximation to assume that the effect of a substituent on the $\mathrm{p}K_a$ of a heteroaromatic system is independent of the number of rings in the molecule. Thus estimates of the $\mathrm{p}K_a$ values of aminoacridines can be made using the approximation:

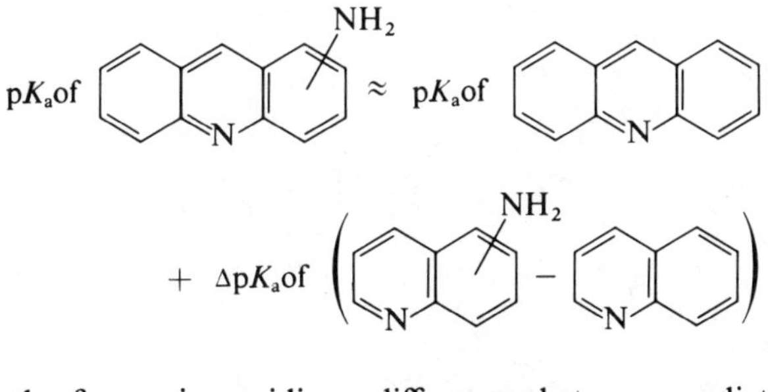

$\mathrm{p}K_a$of [aminoacridine: NH_2, N] $\approx$ $\mathrm{p}K_a$of [acridine: N]

$+\ \Delta\mathrm{p}K_a$of ([aminoquinoline: NH_2, N] $-$ [quinoline: N])

For the five aminoacridines, differences between predicted and experimental values ranged from 0.1 to 0.7, with an average of ± 0.3 (Clark and Perrin, 1964).

7.2 Polyaromatic acids and bases

However, to this point the methods described do not extend to molecules such as 5-aminoquinoline or 6-nitro-l-naphthalene-carboxylic acid where the substituent and the reaction centre are on different rings. One possible approach is that due to Kirkwood and Westheimer (1938; 1939) which deals with the transmission of field effects from substituent to reaction centre through a molecular cavity and the medium, but it is uncertain in its application because of doubt as to what value to assign to the dielectric constant at molecular dimensions.

In an alternative approach, by Dewar and Grisdale (1962), the constant for a substituent varies with the position and is a function of field and mesomeric effects, of the distance between the substituent and the reaction centre, and of formal charge. It is thus possible to convert these σ values for any position into σ_{meta} and σ_{para} values for the ring to which the reaction centre is attached (Perrin, 1965). If a substituent at position j of a polycyclic aromatic molecule acts on a reaction centre at position i,

$$\sigma_{\mathrm{ij}} = F/r_{\mathrm{ij}} + Mq_{\mathrm{ij}} \tag{7.1}$$

where r_{ij} is the distance in benzene C–C bond lengths between positions i and j, and q_{ij} is the electron density at position j produced by attaching $-CH_2^-$ at position i. The parameter F is a measure of the field set up by the substituent, and M is a measure of the combined π-inductive-mesomeric effect of the substituent. An extensive list of q_{ij} values for many common fused-ring systems is given by Longuet-Higgins (1950): as indicated in Table 7.2 they can very easily be obtained from zero order molecular orbitals.

Table 7.2 *Some q_{ij} values*

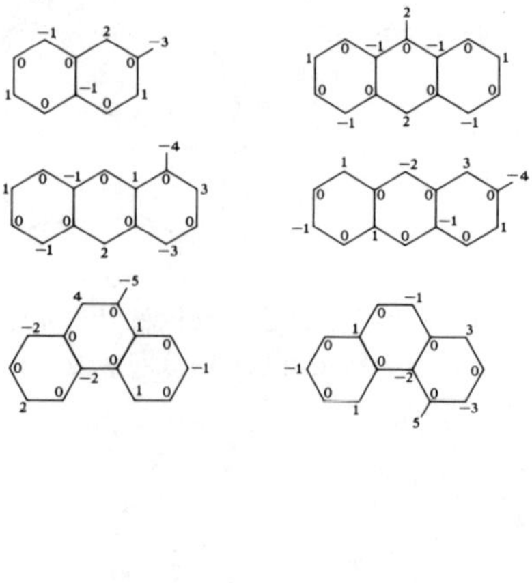

Application to naphthalene will be discussed in detail. If the reactive centre is at position 1 the required distances and the electron densities at the other positions are given by:

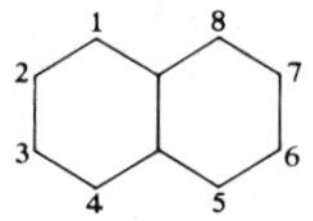

Distances (in benzene C–C bond lengths)

$r_{12} = 1$

$r_{13} = r_{18} = \sqrt{3}$

$r_{14} = 2$

$r_{15} = r_{17} = \sqrt{7}$

$r_{16} = 3$

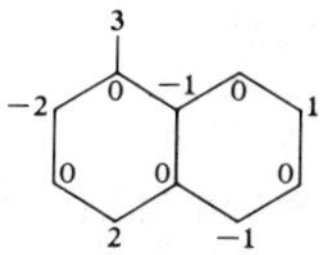

Electron densities

$q_{12} = q_{14} = 4/20† = 1/5$

$q_{13} = q_{16} = q_{18} = 0/20† = 0$

$q_{15} = 1/20†$

For benzene, the corresponding data are:

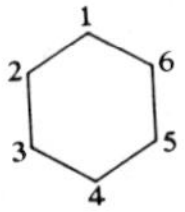

Distances

$r_{12} = r_{16} = 1$

$r_{13} = r_{15} = \sqrt{3}$

$r_{14} = 2$

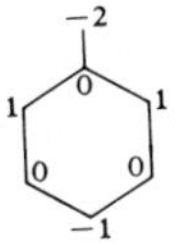

Electron densities

$q_{12} = q_{14} = q_{16} = 1/7‡$

Using Equation (7.1)

$$\sigma_{13} = \sigma_{\text{meta}} = F/\sqrt{3}$$

Therefore, $F = \sqrt{3}\,(\sigma_{\text{meta}}) = 1.732\,\sigma_{\text{meta}}$

$$\sigma_{14} = \sigma_{\text{para}} = 0.50F + M/7$$

Therefore

$$M = 7\sigma_{\text{para}} - 3.50F = 7\sigma_{\text{para}} - 3.50\sqrt{3}(\sigma_{\text{meta}})$$
$$= 7\sigma_{\text{para}} - 6.06\sigma_{\text{meta}}$$

These relations can be applied to polyaromatic molecules by substitution in Equation (7.1) to give an expression for σ_{ij} in terms of the analogous (one ring) benzene derivative. Table 7.3 lists these expressions for substituents in naphthalenes.

† For naphthalene, the normalization coefficient, α^2, is given by $\alpha^2 = 1^2 + 1^2 + 1^2 + 2^2 + 2^2 + 3^2 = 20$.

‡ The normalization coefficient, $\alpha^2 = 1^2 + 1^2 + 1^2 + 2^2 = 7$.

Table 7.3 *Theoretical sigma constants for naphthalene derivatives*

Reaction centre position	Substituent position	Sigma
1	3 or 8	σ_{meta}
1	4	$1.4\sigma_{para} - 0.35\sigma_{meta}$
1	5 or 7	$0.35(\sigma_{meta} + \sigma_{para})$
1	6	$0.58\sigma_{meta}$
2	4	σ_{meta}
2	5	$0.58\sigma_{meta}$
2	6	$(0.13\sigma_{meta} + 0.41\sigma_{para})$†
2	7	$0.50\sigma_{meta}$
2	8	$(0.30\sigma_{meta} + 0.41\sigma_{para})$†

† The expressions given by Perrin (1965) are in error.

Example: 2-Amino-8-hydroxy-3,6-naphthalenedisulfonic acid

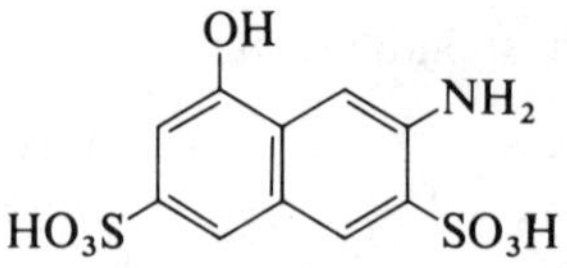

(a) pK_a of $-NH_3^+$

pK_a of a substituted aniline $= 4.58 - 2.88\,\Sigma\sigma$ (Table A.2)

σ_{ortho} for $-SO_3^-$ in position–3 = 0.75 (Table A.5)

σ_{26} for $-SO_3^-$ on position–6
for reaction on position–2
$= 0.13\sigma_{meta}$ for SO_3^- $+ 0.41\sigma_{para}$ for SO_3^- (anilines) (Table 7.3)

$= 0.13\ (0.31)$
$+ 0.41\ (0.46) = 0.23$ (Tables A.1, A.4)

σ_{28} for $-OH$ on position–8
$= 0.30\sigma_{meta} + 0.41\sigma_{para}$ (Table 7.3)
$= 0.30\ (0.13) + 0.41\ (-0.31) = -0.09$ (Tables A.1, A.4)

σ for unsubstituted annelated ring in
β-naphthylamine = 0.08 (Table 7.1)

$\Sigma\sigma$ = 0.97
Predicted pK_a = 1.79
Experimental pK_a = 2.49

(b) pK_a of $-OH$

pK_a of a phenol $= 9.92 - 2.23\,\Sigma\sigma$ (Table A.2)

σ_{meta} for $-SO_3^-$ on the phenolic ring $= 0.31$

σ_{16} for $-SO_3^-$ on position–6 for reaction on position–1
$= 0.58\,\sigma_{meta} = 0.58\,(0.31) = 0.18$ (Tables 7.3, A.1)

σ_{17} for $-NH_2$ on position-7 $= 0.35\sigma_{meta} + 0.35\sigma_{para}$ (Table 7.3)
$= 0.35\ (0.00)$
$+ 0.35\ (-0.29)$ (Tables A.1, A.4)
$= -0.10$

σ for unsubstituted annelated ring in α-naphthol $= 0.28$ (Table 7.1)
$\Sigma\sigma$ $= 0.67$
Predicted pK_a $= 8.43$
Experimental pK_a $= 8.54$

A comparison of predicted and experimental pK_a values of 1- and 2-naphthylamines bearing a substituent at positions 5-, 6-, 7- or 8- showed an average agreement within $\pm 0.2\,pK$ units (Perrin, 1965). Agreement was worst (difference of 0.89) for 8-nitro-1-naphthylamine. If $8-NH_2$, $8-OH$, $8-COO^-$ and $8-NO_2$ were excluded, pK_a values of about 50 quinolines substituted at positions 5-, 6-, 7- and 8- gave predicted pK_a values to within about ± 0.4 pK unit (Perrin, 1965).

Assignment of σ values to ring-bound nitrogens and other heteroatoms greatly extends the range of the method but also increases the uncertainty of any pK_a prediction. For example, consider 6-methylthiopurine, using the Hammett equation for the

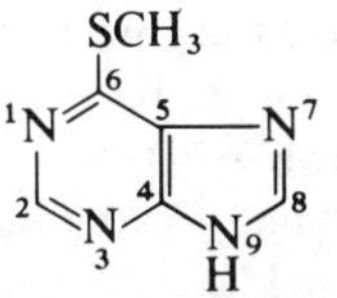

weakly acidic azoles: $pK_a(-NH-) = 17.0 - 4.28\,\Sigma\sigma$. The influence of the ring nitrogens of the annelated six-membered ring and the $-SCH_3$ on $pK_a(-NH-)$ can be approximated by reference to a model compound which treates the purine as though it were a naphthalene derivative so that the data of Table 7.3 can be used. The model, with appropriate renumbering is:

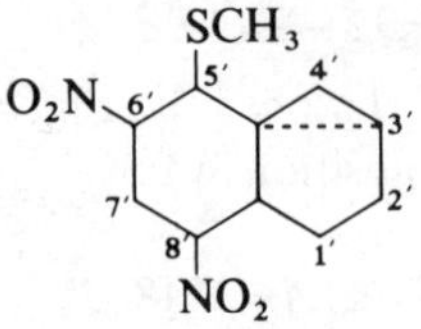

SCH$_3$, O$_2$N, NO$_2$ (structure with positions 1′–8′)

Then

$$\sigma_{1'5'} \text{ for } 5'-SCH_3 = 0.35\,(\sigma_{meta}+\sigma_{para}) \quad \text{(Table 7.3)}$$
$$= 0.35\,(0.14+0.00) \quad \text{(Table A.1)}$$
$$= 0.05$$
$$\sigma_{1'6'} \text{ for } =N- \text{ at position–}6' = 0.58\,\sigma_{meta} \quad \text{(Table 7.3)}$$
$$= 0.58\,(0.73) \quad \text{(Table A.6)}$$
$$= 0.42$$
$$\sigma_{1'8'} \text{ for } =N- \text{ at position–}8' = \sigma_{meta} = 0.73$$
$$\sigma \text{ for } =N- \text{ at position–}3' = \sigma_{meta} = 0.73$$

σ for the annelated benzene ring (quinoline : pyridine)

$$= 0.06 \quad \text{(Table 7.1)}$$
$$\Sigma\sigma = 1.99$$

Then

$$\text{Predicted p}K_a(-NH-) = 17.0 - 4.28\,(1.99)$$
$$= 8.5$$
$$\text{Experimental p}K_a = 8.76$$

However, it is seldom necessary in practice to go back to the parent hydrocarbon and calculate the effects of all substituents *de novo*. Thus, if the pK_a of 4-amino-7-chloroquinoline is required, a logical starting point is the known pK_a (=9.13) of 4-aminoquinoline. Protonation takes place on the ring nitrogen, so the effect of a 7-chloro- on quinoline should be similar to its effect on 4-aminoquinoline. From Table 7.3, σ_{17} for chloro $= 0.35(\sigma_m+\sigma_p)$ $= 0.21$ and ρ for quinoline $= 5.90$ (Table A.2), so that $-\Delta pK_a$ $= 1.24$, and the predicted p$K_a = 7.89$. The experimental value is 8.23.

Similarly, to predict the pK_a of 2-amino-4-methylbenzo-[h]quinoline, the reference compound is benzo[h]quinoline (pK_a $= 4.21$), modified by σ_{ortho} for $-NH_2$ $(= -0.27)$ and σ_{para} for $-CH_3$ $(= -0.14)$. Once again $\rho = 5.90$, so that $-\Delta pK_a = -2.42$, and the predicted p$K_a = 6.63$. The experimental value is 6.70. These are among 24 examples given by Perrin (1965).

The method becomes tedious if 5-membered rings are present.

7.3 Prediction when several acidic or basic centres are present

Sulphonic acids are stronger than carboxylic acids and these, in turn, are usually stronger than phenols. Where several acidic sites are present in a molecule, it is usually necessary to calculate the various sets of pK_a values and to select the set for which the first pK_a is the smallest (strongest acid). Where there are several acidic or basic centres, examination of the pK_a values of model compounds can suggest the likely sequence in which sites in a molecule are protonated or deprotonated. This is helpful, particularly with polyazaheterocyclic compounds such as the purines and pteridines.

In deciding the order of protonation of organic bases, it may be possible to argue by analogy with simpler bases. Thus in 1,6-naphthyridine (*7.7*) the 6-nitrogen would be expected to be more basic

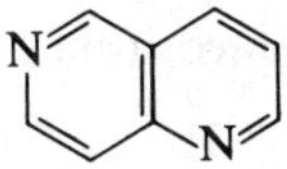

(*7.7*)

than the 1-nitrogen, just as isoquinoline ($pK_a = 5.42$) is a stronger base than quinoline ($pK_a = 4.92$) or 5-nitro-2-azanaphthylamine ($pK_a = 3.49$) is stronger than 6-nitro-1-azanaphthalene ($pK_a = 2.92$).

The principles involved can be illustrated by considering the pK_a values of 3-(2-aminoethylaminomethyl)pyridine (*7.8*). Inspection shows three nitrogens to be present in the molecule, namely a primary amino ($pK_a \sim 10.8$), a secondary amino ($pK_a \sim 11.2$) and a pyridine nitrogen ($pK_a \sim 5.2$). The site for addition of the first and second protons (strongest bases) is therefore likely to be the aliphatic nitrogens.

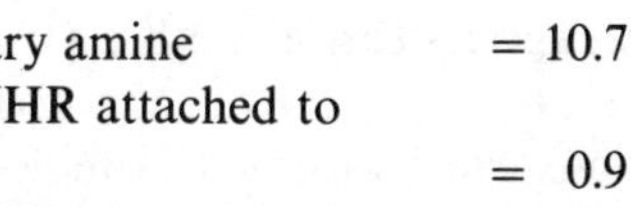

$CH_2NHCH_2CH_2NH_2$

(*7.8*)

For protonation on the terminal $-NH_2$:

pK_a of a primary amine	$= 10.77$	
$-\Delta pK_a$ for -NHR attached to β-carbon	$= 0.9$	(Table 4.2)
$-\Delta pK_a$ for 3-pyridyl removed by four atoms from the basic centre $= (0.4)^3 \times 2.7$	$= 0.17$	(Table 4.2)
Predicted pK_a	$= 9.70$	

For protonation on the secondary nitrogen:

pK_a of a secondary amine	= 11.15
$-\Delta pK_a$ for $-NH_2$ attached to β-carbon	= 0.8
$-\Delta pK_a$ for 3-pyridyl attached to α-carbon	= 2.7
Predicted pK_a	= 7.65

Hence protonation on the terminal nitrogen (higher pK_a) is the preferred site. The experimental pK_a for the first protonation is 9.64.

For the third protonation on the pyridine nitrogen, the aliphatic nitrogens are already protonated:

pK_a of a substituted pyridine $= 5.25 - 5.90\,\Sigma\sigma$ (Table A.3)

σ_{meta} for the substituent, $-CH_2\overset{+}{N}H_2CH_2CH_2NH_3^+$, is approximated by σ_{meta} for $-CH_2\overset{+}{N}H_2CH_2CH_2CH_2CH_3 + \sigma_{meta}$ for $-(CH_2)_4NH_3^+$ (both of which can be estimated from data in Table A.1)

$= 0.4 \times \sigma_{meta}$ for $-NH_2CH_2CH_2CH_2CH_3^+ \quad + 0.4 \times 0.4 \times$ σ_{meta} for $CH_2CH_2NH_3^+$

$= 0.4 \times 0.71 + 0.4 \times 0.4 \times 0.23$

$= 0.32$

Predicted pK_a $= 3.36$

Experimental pK_a $= 3.31$

Pyrimidine (pK_a = 1.3) is a much weaker base than imidazole (pK_a = 6.95) or benzimidazole (pK_a = 5.53) so that when the imidazole and pyrimidine rings are fused to give purine, the stronger basic centres would be expected to reside on the imidazole ring. That this is so can be seen if the pK_a values of substituted purines are plotted against the pK_a values of the corresponding benzimidazoles: a straight line having a slope of unity and displaced downwards by 2.9 ± 0.2 pH units can be drawn through the plotted points. The vertical displacement downwards measures the effect of the nitrogen atoms in the benzene ring. The most basic nitrogen is $N_{(7)}$ or $N_{(9)}$, so that for pK_a predictions purines can be looked on as substituted imidazoles.

Where the neutral species carries both a positive and a negative charge, such as $\overset{+}{N}H_3-CH_2-COO^-$, the species is known as a zwitterion. Zwitterions can be formed in molecules which contain both acidic and basic centres if the pK_a of the protonated base (for example, $-NH_3^+$) is close to, or greater than, pK_a of the acid group yielding the anion (for example, $-COOH$). Consider, for example, glutamic acid. Predictions concerning its ionization behaviour can begin with the fully protonated species (*7.9*) which exists in strongly acid solution. The predicted pK_a for proton removal from the carbon-5 carboxyl is calculated as in Section 4.2:

$$\mathrm{HOOC{-}CH_2{-}CH_2{-}\underset{}{\overset{\overset{\displaystyle NH_3^+}{|}}{C}}H{-}COOH}$$

(*7.9*)

$$pK_a = 4.80 - 0.39 - 0.22 = 4.19$$

(4.80 = pK_a of a simple aliphatic carboxylic acid; $0.39 = (0.4)^2 \times 2.43 = -\Delta pK_a$ for $-NH_3^+$ on a γ-carbon; $0.22 = (0.4)^2 \times 1.37 = -\Delta pK_a$ for -COOH on a γ-carbon).

Alternatively, the predicted pK_a for proton removal from the carbon-1 carboxyl is:

$$pK_a = 4.80 - 2.43 - 0.22 = 2.15$$

(4.80 = pK_a of a simple aliphatic carboxylic acid; 2.43 = $-\Delta pK_a$ for $-NH_3^+$ on an α-carbon; 0.22 = $-\Delta pK_a$ for $-COOH$ on a γ-carbon).

If the first proton removed comes from $-NH_3^+$, the predicted pK_a would be:

$$pK_a = 10.77 - 3.0 - 0.48 = 7.29$$

(10.77 = pK_a of a protonated primary aliphatic amine; 3.0 = $-\Delta pK_a$ for $-COOR$† on an α-carbon; 0.48 = $-\Delta pK_a$ for $-COOR$† on a γ-carbon). Hence the first proton loss is from the carbon-1 carboxyl with predicted $pK_a = 2.15$. The second pK_a is for the carbon-5 carboxyl for which the predicted pK_a now becomes:

$$pK_a = 4.80 + 0.10 - 0.39 = 4.51$$

(4.80 = pK_a of a simple aliphatic carboxylic acid; $-0.10 = -\Delta pK_a$ for $-COO^-$ on a γ-carbon; 0.39 = $-\Delta pK_a$ for $-NH_3^+$ on a γ-carbon).

The third pK_a is for $-NH_3^+$ for which:

$$\text{predicted } pK_a = 10.77 - 0.8 + 0.08 = 10.05$$

(10.77 = pK_a for a protonated primary aliphatic amine; -0.8‡ =

† The value for $-COOH$ is not available.
‡ See footnote ∥ Table 4.2.

$-\Delta pK_a$ for $-COO^-$ on an α-carbon; 0.08 = $-\Delta pK_a$ for $-COO^-$ on a γ-carbon).

On the basis of the predicted pK_a values: 2.15 ($-COOH$), 4.51 ($-COOH$), 10.05 ($-NH_3^+$), glutamic acid would be predominantly zwitterionic in the pH range, (2.15 + 2) to (10.05 − 2). The experimental pK_a values are 2.13, 4.32 and 9.76.

As a further example, consider 3-aminophenol. If it is assumed that the phenolic pK_a is less than the aminium pK_a as required for zwitterion formation, the predicted values are:

$$pK_{a1}(-OH) = 9.92 - 2.23\,\sigma_{meta} \text{ for } -NH_3^+ = 8.43$$

and

$$pK_{a2}(-NH_3^+) = 4.58 - 2.88\,\sigma_{meta} \text{ for } -O^- = 5.93$$

which are inconsistent with zwitterion formation. However, for the assumption that pK_a ($-OH$) is greater than pK_a ($-NH_3^+$), the predicted values are:

$$pK_{a1}(-NH_3^+) = 4.58 - 2.88\,\sigma_{meta} \text{ for } -OH = 4.21$$

and

$$pK_{a2}(-OH) = 9.92 - 2.23\,\sigma_{meta} \text{ for } -NH_2 = 9.92$$

which approximate the experimental values, 4.37 and 9.82. Hence it can be concluded from the predicted values that 3-aminophenol is not zwitterionic in aqueous solution.

7.4 Prediction by considering fragments of molecules

Especially among natural products, there are large molecules having only a small number of acidic or basic centres. It is sometimes possible, by taking account of the attenuation of effects along a saturated aliphatic chain or an alicyclic ring, to decrease the size of the molecule that needs to be considered or to resolve the molecule into several smaller components, each containing only one acidic or basic centre. Thus the three pK_a values (two phenolic, one secondary amine) of tubocurarine can be predicted from three separate fragments (Perrin, 1980). Quinine was also conveniently divided into a quinoline and a non-aromatic portion (Perrin, 1980).

7.4.1 Hyoscyamine

At its simplest level there is the comparison of hyoscyamine (*7.10*) ($pK_a = 9.7$) with diethylmethylamine to yield a predicted pK_a of 10.5 (based on $pK_a = 10.5$ for a tertiary amine, less 0.2 for the methyl bound to the nitrogen plus 0.2 for ring formation involving the nitrogen). Agreement is improved if allowance is made for the effect of the $-O.CO.R$ group in hyoscyamine (subtraction of 2×0.68† for the effect of this group transmitted along two chains of three carbon atoms), giving a predicted pK_a of 9.1

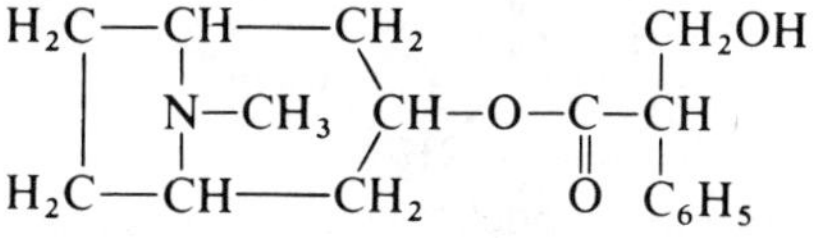

(*7.10*)

7.4.2 Novobiocin

For the purposes of pK_a prediction, the novobiocin molecule can be divided into three portions as shown in (*7.11*):

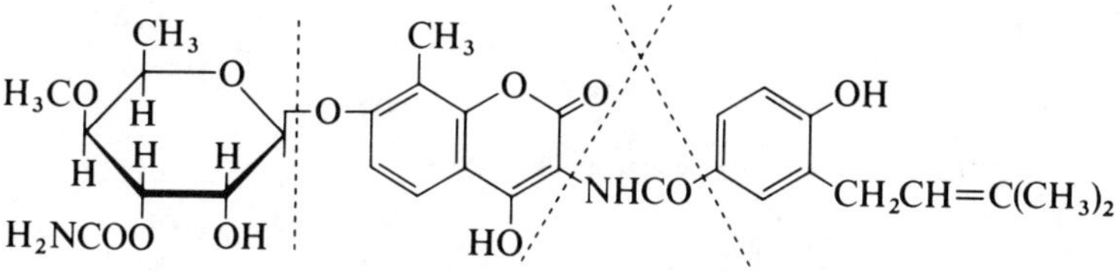

(*7.11*)

Fragment I is approximated by:

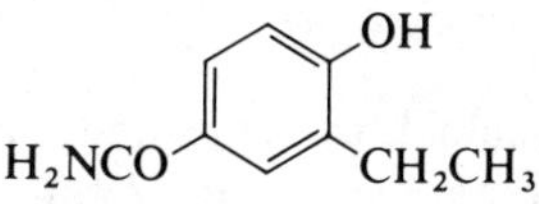

pK_a of a phenol $= 9.92 - 2.23\,\Sigma\sigma$
σ_{ortho} for $-C_2H_5$ $= -0.09$
σ_{para} for $-CONH_2$ $= 0.31$
Predicted pK_a $= 9.43$
Experimental pK_a $= 9.1$

† 1.7 (from Table 4.2) × 0.4

Fragment II is approximated by:

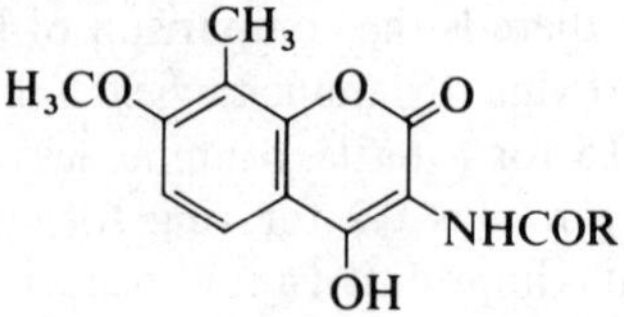

$$H_3CO,\ CH_3 \text{-substituted 4-hydroxy-3-NHCOR-coumarin}$$

for which,

$$\text{3-acetyl-4-hydroxycoumarin (}COCH_3\text{, OH)}$$

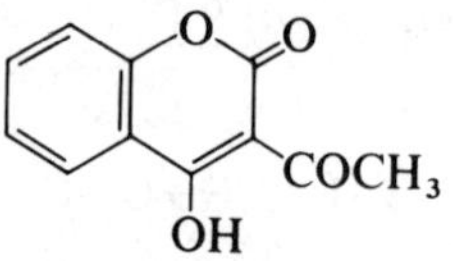

with pK_a = 4.3 serves as a model, taking $-COCH_3$ to approximate to $-NHCOR$. The effects of $-OCH_3$ and $-CH_3$ in the unsubstituted ring in the model can be estimated using the Dewar–Grisdale method (Section 7.2), taking the benzpyran system to approximate to naphthalene:

σ for $-CH_3$ on position–5 $= 0.35\ (\sigma_{meta} + \sigma_{para})$ (Table 7.3)

$= 0.35\ (-0.06 - 0.14) = -0.07$

σ for $-OCH_3$ on position–6 $= 0.58\ \sigma_{meta}$ (Table 7.3)

$= 0.58\ (0.11) = 0.06$

Thus, the effects are self-cancelling and the predicted p$K_a \sim 4.3$. The experimental value is 4.2.

Fragment III has no appreciably acidic or basic centres.

7.4.3 Narceine

The structure of the alkaloid, narceine, and a convenient fragmentation for pK_a prediction are shown in (*7.12*).

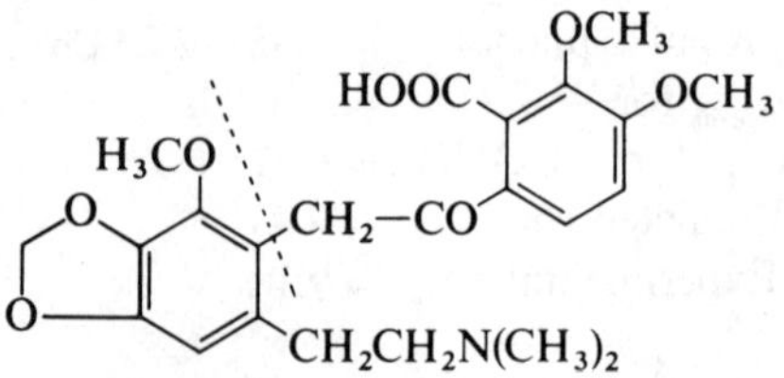

$$\text{OCH}_3,\ \text{HOOC},\ \text{OCH}_3,\ H_3CO,\ CH_2{-}CO,\ CH_2CH_2N(CH_3)_2$$

(*7.12*)

Fragment I approximates to $C_6H_5CH_2CH_2N(CH_3)_2$ for which:

$$\text{predicted p}K_a = 10.5 - 0.4 - 0.8 = 9.3$$

(10.5 for a tertiary amine; -0.4 for two $N-CH_3$ groups; -0.8 for C_6H_5- on a β-carbon). The experimental value is 9.3.

Fragment II approximates to

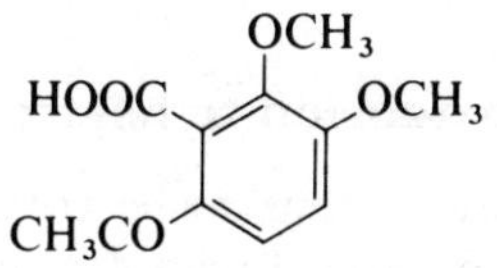

OCH_3
HOOC OCH_3
CH_3CO

for which

$$\text{predicted p}K_a = 4.20 - \sigma_{ortho} \text{ for } -OCH_3 - \sigma_{meta} \text{ for} -OCH_3 - \sigma_{ortho} \text{ for } COCH_3$$

$$= 4.20 - 0.12 - 0.11 - 0.07 = 3.90$$

Experimental value is 3.3.

Chapter Eight

Extension of the Hammett and Taft Equations

8.1 An empirical extension to heterocycles

Hammett and Taft-type equations can be written for the ionization of many types of acids and bases. Tables A.2 and A.3 list some of the published examples. It has been found empirically (Jaffé, 1953) that as a working approximation the hetero group (–O–, –S–, –NH–) in an unsaturated 5-membered ring carrying an acidic or basic group can be replaced by –C(–OR, –SR, –NHR) = CH–, respectively, giving a benzene derivative. An acidic or basic substituent in the 2- (or 3-) position with respect to the hetero-group remains in the 2- (or 3-) position with respect to the –OR, –SR, –NHR in the benzene derivative. Transmission of electronic effects across the heteroatom in the 5-membered ring is not large, so that the substituents on the 5-membered ring can be approximated by substituents on a 6-membered ring as shown:

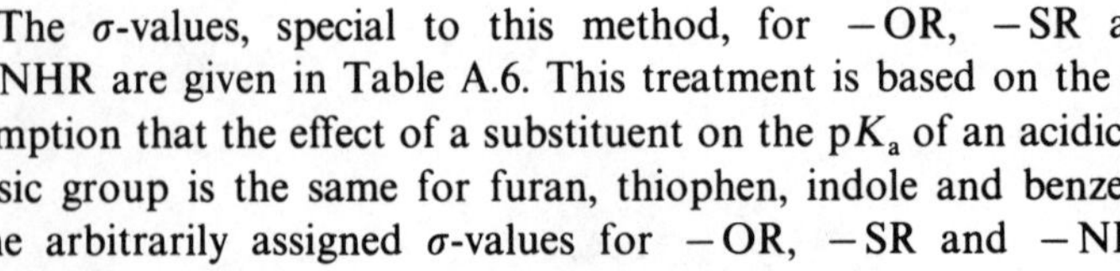

The σ-values, special to this method, for –OR, –SR and –NHR are given in Table A.6. This treatment is based on the assumption that the effect of a substituent on the pK_a of an acidic or basic group is the same for furan, thiophen, indole and benzene. The arbitrarily assigned σ-values for –OR, –SR and –NHR set the scale from which to measure changes in pK_a produced by adding the substituent. The following examples illustrate the method.

8.1.1 4-Nitrothiophen-2-carboxylic acid

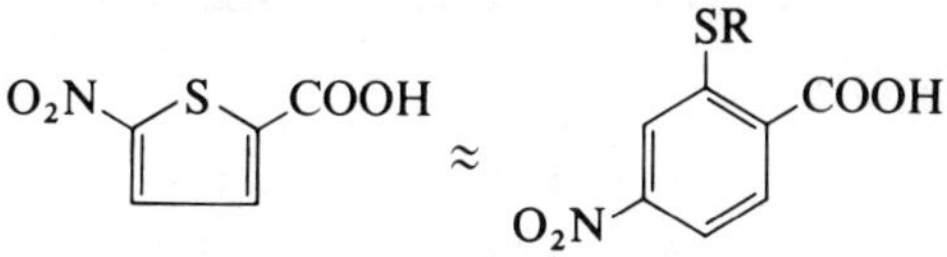

O_2N–thiophene–COOH ≈ 2-SR-5-O_2N-benzoic acid

Predicted pK_a for an aromatic carboxylic acid	$= 4.20 - 1.00\,\Sigma\sigma$	(Table A.2)
σ_{ortho} for $-SR$	$= 0.72$	(Table A.6)
σ_{para} for $-NO_2$	$= 0.78$	(Table A.1)
Predicted pK_a	$= 4.20 - (0.72 + 0.78)$	
	$= 2.70$	
Experimental pK_a	$= 2.68$	

8.1.2 4-Bromopyrrole-2-carboxylic acid

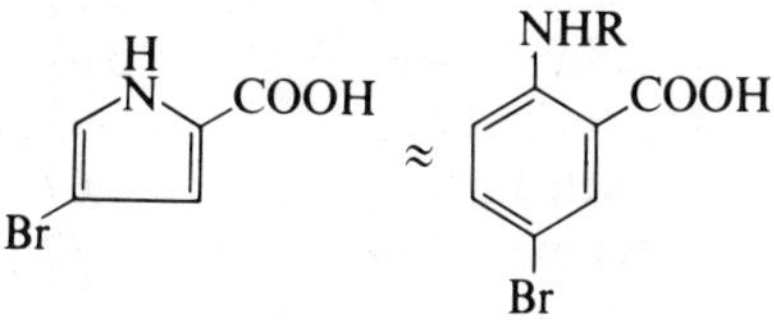

Br–pyrrole(NH)–COOH ≈ 2-NHR-5-Br-benzoic acid

As in 8.1.1, predicted pK_a	$= 4.20 - 1.00\,\Sigma\sigma$	
σ_{ortho} for $-NHR$	$= -0.24$	(Table A.6)
σ_{meta} for $-Br$	$= 0.39$	
Predicted pK_a	$= 4.05$	
Experimental pK_a	$= 4.06$	

8.1.3 Furan-2,4-dicarboxylic acid

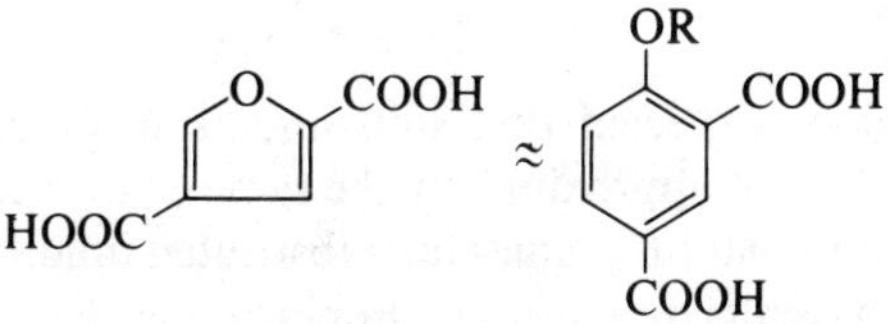

HOOC–furan–COOH ≈ 2-OR-benzene-1,4-dicarboxylic acid (COOH, OR, COOH)

As in Section 8.1.1, predicted p$K_a = 4.20 - 1.00\,\Sigma\sigma$. p$K_a$ for $-COOH-2$:

σ_{ortho} for $-OR$	$= 1.08$	(Table A.6)
σ_{meta} for $-COOH$	$= 0.35$	(Table A.1)

Predicted pK_a	$= 2.77$	
Experimental pK_a	$= 2.63$	

pK_a for $-COOH-4$:

σ_{meta} for $-OR$	$= 0.25$	(Table A.6)
σ_{meta} for $-COO^-$	$= 0.09$	(Table A.1)
Predicted pK_a	$= 3.86$	
Experimental pK_a	$= 3.77$	

8.1.4 5-Methylselenofuran-2-carboxylic acid

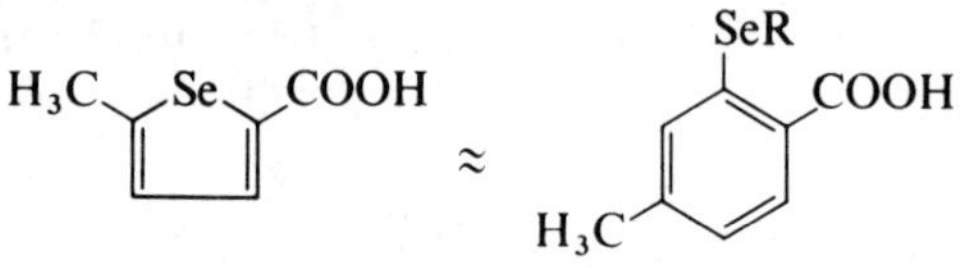

As in Section 8.1.1, predicted p$K_a = 4.20 - 1.00\ \Sigma\sigma$.

σ_{ortho} for $-SeR$	$= 0.67$	(Table A.6)
σ_{para} for $-CH_3$	$= -0.14$	(Table A.1)
Predicted pK_a	$= 3.67$	
Experimental pK_a	$= 3.82$	

The base, imidazole, protonates on $=N-$, so that the notional replacement of the heterogroup, $-NH-$ by $-C(NHR){=}CH-$ leads to a pyridine substituted in the 2-position:

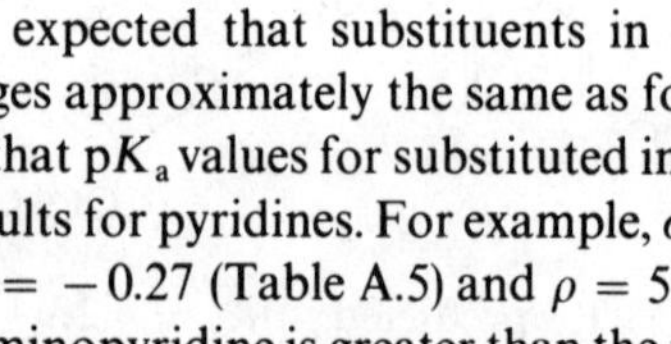

Thus it would be expected that substituents in imidazoles would produce pK_a changes approximately the same as for the corresponding pyridines, and that pK_a values for substituted imidazoles could be predicted using results for pyridines. For example, σ_{ortho} for $-NH_2$ in the pyridine series $= -0.27$ (Table A.5) and $\rho = 5.90$ (Table A.3), so that the pK_a of 2-aminopyridine is greater than the pK_a of pyridine by 1.59. Thus the predicted pK_a of 2-aminoimidazole at 25° = the pK_a of imidazole at 25° $(= 6.99) + 1.59 = 8.58$; the experimental value is 8.46. This procedure leads to the general expectation that 2- and 4-substituted imidazoles (cf., 2- and 6-substituted pyridines) have approximately the same pK_a values. The pK_a value of 2-methylimidazole and 4-methylimidazole are 7.86 and 7.52, respectively; likewise, 2-phenylimidazole (= 6.44) and 4-phenylimidazole (= 6.05).

Substituted pyrazoles have pK_a values near to those predicted by an argument similar to that for imidazole, but in general, predictions are less reliable. For example, see the following.

8.1.5 Pyrazole-3-carboxylic acid

H N N COOH ≈ NHR N COOH

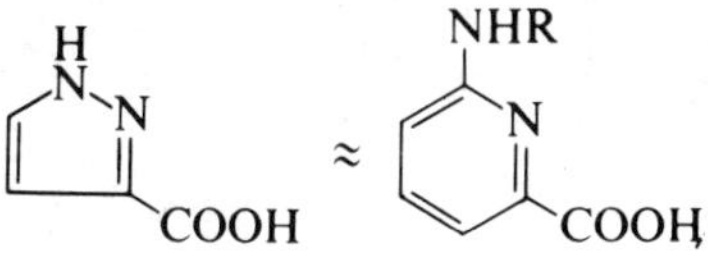

As in Section 8.1.1, pK_a (for $-COOH$) $= 4.20 - 1.00\,\Sigma\sigma$.

σ_{ortho} for $=N-$	$= 0.56$	(Table A.6)
σ_{meta} for $-NHR$	$= -0.34$	(Table A.6)
Predicted pK_a	$= 3.98$	
Experimental pK_a	$= 3.74$	

(The pK_a values of the protonated basic centres in pyrazoles are about 2 and hence zwitterion formation does not need to be considered).

Similarly, the predicted pK_a for 5-methylpyrazole-3-carboxylic acid (σ_{meta} for $-CH_3 = -0.06$) = 4.04; the experimental value = 3.79.

The method can be extended to compounds with a greater number of ring nitrogens.

8.1.6 1,2,3-Triazole-4-carboxylic acid

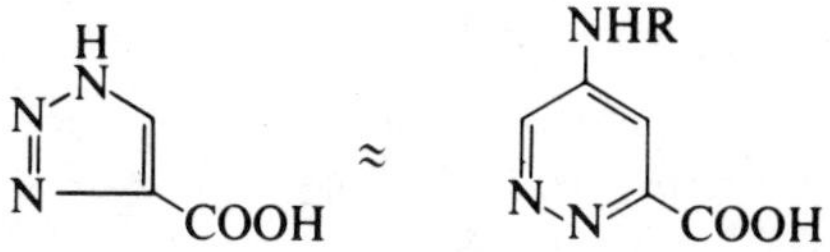

H N N N COOH ≈ NHR N N COOH

As in Section 8.1.1, pK_a ($-COOH$)	$= 4.20 - 1.00\,\Sigma\sigma$	
σ_{meta} for $-NHR$	$= -0.34$	(Table A.6)
σ_{ortho} for ($-N=$)	$= 0.73$	(Table A.6)
σ_{meta} for ($-N=$)	$= 0.56$	(Table A.6)

Predicted pK_a = 3.25
Experimental pK_a = 3.22

The procedure can also be extended to annelated rings so that

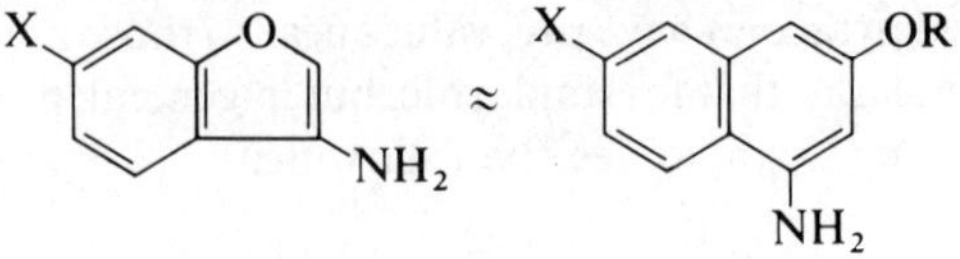

and

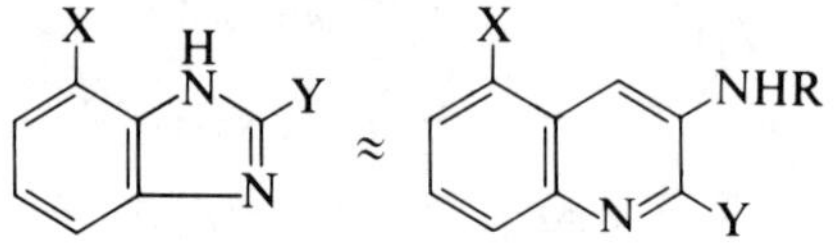

In addition to Hammett equations for classical aromatic systems, Hammett-type equations can be written for other systems containing conjugated double bonds, both in open chains and rings. For example, Table A.2 gives Hammett-type equations for the pK_a values of *trans*-3-substituted acrylic acids (Charton and Meislich, 1958):

$$pK_a = 4.39 - 2.23\,\sigma_{para} \quad (8.1)$$

and for 5-substituted tropolones (*8.1*) (Oka *et al.*, 1962):

$$pK_a = 6.42 - 3.10\,\sigma_{para} \quad (8.2)$$

where the σ_{para} constants (Table A.4) are special to phenols.

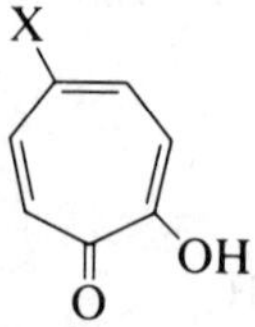

(*8.1*)

Equation (8.2) can also be used to predict the pK_a of tropolones substituted on position -3 and -4 by regarding these as equivalent to *ortho* and *meta* positions, respectively, in phenols.

8.1.7 3-Bromohinokiol

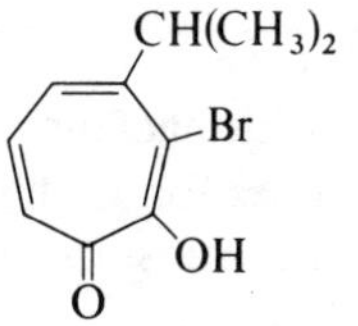

$CH(CH_3)_2$
Br
OH
O

Predicted pK_a	$= 6.42 - 3.10\ (\sigma_{ortho}$ for $-Br + \sigma_{meta}$ for $-CH(CH_3)_2)$
	$= 4.5$
Experimental pK_a	$= 4.68$

8.2 The prediction of ρ values and generalized Hammett and Taft equations

Inspection of the set of Hammett ρ values for proton addition or removal shows that there is a strong correlation between the value of ρ and i, the number of atoms by which the ionizable hydrogen is separated from the aromatic ring (Barlin and Perrin, 1966). The relation is

$$\rho = (2.4)^{2-i} \tag{8.3}$$

based on experimental values ranging from $\rho = 5.90$ $(i = 0)$ for proton addition to pyridine to $\rho = 0.21$ $(i = 4)$ for proton removal from phenylpropionic acids. Equation (8.3) gives: $i = 0$, $\rho = 5.76$; $i = 1$, $\rho = 2.4$; $i = 2$, $\rho = 1$; $i = 3$, $\rho = 0.42$; $i = 4$, $\rho = 0.17$.

Thus it is possible to hazard a reasonable guess at the pK_a of an acid or base of an unknown series if the pK_a of one of its members is known. The ρ value fixes the slope of the relationship and extrapolation to the first member of the series gives the constant term, pK_a°. For thiobenzoic acid, $pK_a = 2.61$ and $i = 2$ so that the pK_a values of other members of this series would be expected to be given to within a few tenths of a pH unit by

$$pK_a = 2.61 - 1.0\,\Sigma\sigma \tag{8.4}$$

where $\rho = 1.0$ as predicted and as for benzoic acid.

For thiophenols, the predicted equation would be

$$pK_a = 6.62\dagger - 2.4\ddagger\,\Sigma\sigma$$

† 6.62 = the pK_a of thiophenol
‡ $i = 1$

A better fit is obtained by the equation

$$pK_a = 6.62 - 2.2\,\Sigma\sigma \tag{8.5}$$

where $\rho = 2.2$ by analogy with phenols.

Similarly, the Taft ρ^* values for aliphatic systems depend mainly on the distance between the substituent and the reaction centre, with

$$\rho^* \sim 0.8 \times (2.0)^{2-i} \tag{8.6}$$

making it possible to write Taft equations given the pK_a of at least one member of a series of acids or bases. In this relation i is the number of atoms in a saturated chain between the substituent and the atom to which the ionizable hydrogen is attached. Equation (8.6) gives: $i = 0$, $\rho^* = 3.2$; $i = 1$, $\rho^* = 1.6$; $i = 2$, $\rho^* = 0.8$; $i = 3$, $\rho^* = 0.4$.

Note that for acetic acids, RCH_2COOH, $i = 2$ giving a predicted $\rho^* = 0.8$; the experimental Taft equation for RCH_2COOH gives $\rho^* = 0.67$ from $pK_a = 4.76 - 0.67\,\sigma^*$. Similarly, for methanoic acids, predicted $\rho^* = 1.6$, experimental $\rho^* = 1.62$.

If the effect is transmitted over a double or triple bond, the effective chain length is decreased and ρ^* is increased. Thus for enolic forms of substituted acetylacetones:

$$CH_3-CO-\underset{}{\overset{\displaystyle R}{\overset{|}{C}}}=\overset{\displaystyle CH_3}{\overset{|}{C}}-OH \rightleftarrows CH_3-CO-\overset{\displaystyle R}{\overset{|}{C}}=\overset{\displaystyle CH_3}{\overset{|}{C}}-O^- + H^+$$

i is effectively less than 2 so that ρ^* would be expected to be greater than 0.8. In fact, ρ^* is 1.78 [from $pK_a = 9.25 - 1.78\,\sigma^*$ (Table A.2)] so that i is effectively about 1. Similarly, for 3-substituted-2-hydroxy-1,4-naphthoquinones, which contain the group, $R-C=C-OH$, $\rho^* = 1.40$ [$pK_a = 5.16 - 1.40\,\sigma^*$ (Table A.2)].

The pK_a of thioacetic acid is 1.14 pH units less than for the acetic acid: hence, from the foregoing discussion, the expected relationship of aliphatic thiocarboxylic acids with their oxygen analogues is:

$$(pK_a)_{RCOSH} = (pK_a)_{RCOOH} - 1.14 \tag{8.7}$$

Applications of the method are shown in the following examples.

8.2.1 3-Mercaptobenzoic acid

Predicted pK_a for ArSH $= 6.62 - 2.2\,(\sigma_{meta}$ for $-COO^-$†) (Equation 8.5)
$= 6.62 - 2.2\,(0.09) = 6.42$
Experimental value $= 6.32$

8.2.2 2-Hydroxy-3-methylbutanamidine

$$H_3C-\underset{\displaystyle CH_3}{\underset{|}{CH}}-CH(OH)-C(=NH)NH_2$$

A Taft equation can be postulated for protonated amidines, $RC(=NH)NH_3^+$, from the experimental pK_a of $CH_3C(=NH)NH_3^+$ ($= 12.1$) and the predicted ρ^* for $i = 1$:

$$pK_a = 12.1 - 1.6\sigma^* \tag{8.8}$$

σ^* for $(CH_3)_2CHCH(OH)-$ can be estimated as described in Chapter 4, Relation 4.6:

$$\begin{aligned}\sigma^* \text{ for } (CH_3)_2CHCH(OH)- &\sim \sigma^* \text{ for } (CH_3)_2CHCH_2- \\ &\quad + \sigma^* \text{ for } CH_3 \\ &\quad + \sigma^* \text{ for } CH_2OH \\ &\sim -0.19 + 0 + 0.62 \sim 0.43\end{aligned}$$

So $pK_a \sim 12.1 - 1.6(0.43) \sim 11.4$

Experimental value $= 11.35$

8.3 Cation–Pseudobase Equilibria

Pseudobase formation results from the covalent addition of a water molecule to an unsaturated heterocyclic cation in aqueous solution with the release of a proton:

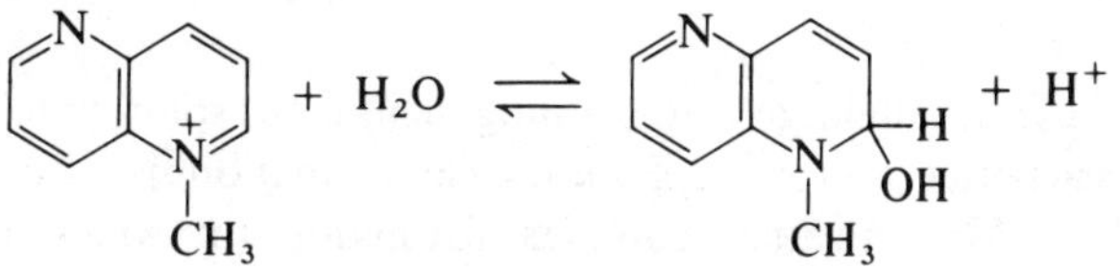

N, $\overset{+}{N}$–CH_3 + H_2O $\rightleftharpoons$ N, N–CH_3, H, OH + H^+

† Predicted pK_a for ArCOOH $= 4.20 - \sigma_{meta}$ for $-SH = 4.20 - 0.25 = 3.95$.

The equilibrium constant for pseudobase formation can be written as:

$$K_{R^+} = [H^+][QOH]/[Q^+]$$

which is analogous to the pK_a of a Bronsted acid and is the hydrogen ion concentration at which QOH and Q^+ are present in equal amounts.

Pseudobase formation can occur in many nitrogen heterocycles (see Bunting, 1979) and the effects of substituents can be represented by Hammett equations. Examples are given in Table 8.1. The

Table 8.1 *Hammett and Taft equations for some cations forming pseudobases–(Bunting, 1979).*

Cation	
1-X-3-Methyl-2-oxopyrimidinium	$7.1 - 3.60\sigma^*$
1-(X-Benzyl)-5-nitroquinolinium	$11.38 - 1.32\sigma$
2-X-5-Nitroisoquinolinium	$11.6 - 3.70\sigma^*$
2-(X-Benzyl)-5-nitroisoquinolinium	$11.29 - 1.14\sigma$
1-X-1, 8-Naphthyridium	$12.5 - 4.90\sigma^*$
1-Cyano-6-X-quinolinium	$-1.21 - 6.16\sigma_{para}$
1-Cyano-5-X-quinolinium	$-1.0 - 5.8(0.5\sigma^+_{para} + 0.5\sigma^+_{meta})$
1-Cyano-7-X-quinolinium	$-1.1 - 6.6(0.8\sigma_{para} + 0.2\sigma^+_{para})$
6-X-Benzopyrilium	$-2.05 - 5.64\sigma_{para}$
7-X-Benzopyrilium	$-1.82 - 5.90\sigma_{para}$
6-X-Benzothiopyrilium	$3.17 - 5.06\sigma_{para}$
7-X-Benzothiopyrilium	$3.25 - 6.09\sigma_{para}$
6-X-Isobenzothiopyrilium	$2.61 - 6.93\sigma_{para}$
7-X-Isobenzothiopyrilium	$2.16 - 4.88\sigma_{para}$
2-(X-Phenyl)benzothiopyrilium	$5.90 - 2.70\sigma$
4-(X-Phenyl)benzothiopyrilium	$3.58 - 1.85\sigma$
4-(X-Phenyl)-1,3-dithiolium	$2.10 - 1.67\sigma$
3-(X-Phenyl)-1,1-dimethyl -2-phenyl-1*H*-isoindolium	$10.4 - 0.9\sigma^+$
3-(X-Phenyl)-1,1,2-triphenyl-1*H* -isoindolium	$7.3 - 1.17\sigma^*$
3-(X-Phenyl)-1,1-dimethyl-1*H* -isobenzofurylium	$1.52 - 3.75\sigma$

pseudobase can undergo ring-opening in aqueous solution to give an amino-carbonyl tautomer. Pseudobase formation is also closely related to Meisenheimer complex formation (in which there is nucleophilic addition to an electron deficient neutral aromatic molecule).

8.4 Linear free energy relations for pK_a values of organic phosphorus acids

A set of σ^ϕ constants for substituents bonded to phosphorus has been used to correlate the ionization constants of organophosphorus acids (Mastryukova and Kabachnik, 1971). With these σ^ϕ constants (given in Table 8.2), a Hammett equation can be applied, as long as the ionizable proton is not directly attached to phosphorus. The equations that have been obtained are given in Table 8.3. The ionization

Table 8.2 *Special σ^ϕ values for organic phosphorus acids (Mastryukova and Kabachnik, 1969; 1971)*

Substituent	σ^ϕ
C_0; *1 element*	
$-Cl$	0.93
$-F$	0.56
$-H$	0.00
$-O^-$	0
C_0; *2 elements*	
$-OH$	−0.39
C_1; *2 elements*	
$-CCl_3$	0.3
$-CF_3$	0.7
$-CH_3$	−0.96
C_1; *3 elements*	
$-CHCl_2$	0.27
$-CH_2Br$	0.0
$-CH_2Cl$	−0.05
$-CH_2I$	−0.1
$-CH_2OH$	−0.55
$-OCH_3$	−0.12
$-SCH_3$	0.15
C_2; *2 elements*	
$-CH=CH_2$	−0.68
$-C_2H_5$	−1.10
C_2; *3 elements*	
$-CH_2CH_2Br$	−0.8
$-OC_2H_5$	−0.21
$-SC_2H_5$	0.09
$-N(CH_3)_2$	−1.22

Table 8.2 (*Contd.*)

Substituent	σ^{ϕ}
C_2; *4 elements*	
$-OCH_2CH_2Cl$	0.03
C_3; *2 elements*	
$-CH_2CH=CH_2$	−0.83
$-CH(CH_3)CH_3$	−1.30
$-CH_2CH_2CH_3$	−1.18
C_3; *3 elements*	
$-CH_2CH_2CN$	−0.6
$-CH_2CH_2CH_2Cl$	−0.72
$-OCH(CH_3)CH_3$	−0.29
$-OCH_2CH_2CH_3$	−0.32
$-SCH(CH_3)CH_3$	−0.06
$-SCH_2CH_2CH_3$	−0.06
C_4; *2 elements*	
$-(CH_2)_3CH_3$	−1.22
$-CH(CH_3)C_2H_5$	−1.36
$-C(CH_3)_3$	−1.55
$-CH_2CH(CH_3)CH_3$	−1.30
C_4; *3 elements*	
$-O(CH_2)_3CH_3$	−0.41
$-CH_2CH_2OC_2H_5$	−0.77
$-OCH_2CH(CH_3)CH_3$	−0.30
$-N(C_2H_5)_2$	−1.54
$-CH_2Si(CH_3)_3$	−1.6
C_5; *2 elements*	
$-cyclo-C_5H_9$	−1.25
$-(CH_2)_2CH(CH_3)_2$	−1.27
$-CH_2C(CH_3)_3$	−1.44
$-(CH_2)_4CH_3$	−1.21
$-C(CH_3)_2C_2H_5$	−1.54
C_5; *3 elements*	
$-OCH_2CH_2CH(CH_3)CH_3$	−0.38
$-O(CH_2)_4CH_3$	−0.39
$-OCH_2C(CH_3)_3$	−0.29
C_6; *2 elements*	
$-C_6F_5$	−0.02
$-C_6H_5$	−0.48
$-cyclo-C_6H_{11}$	−1.19
$-(CH_2)_5CH_3$	−1.21

Table 8.2 (*Contd.*)

Substituent	σ^{φ}
C_6; *3 elements*	
$-(C_6H_4-3-Br)$	−0.23
$-(C_6H_4-4-Br)$	−0.25
$-(C_6H_4-3-Cl)$	−0.22
$-(C_6H_4-4-Cl)$	−0.29
$-(C_6H_4-3-OH)$	−0.32
$-(C_6H_4-4-OH)$	−0.65
$-OC_6H_5$	−0.06
$-(C_6H_4-3-NH_2)$	−0.56
$-(C_6H_4-4-NH_2)$	−0.78
$-(C_6H_4-3-NHNH_2)$	−0.46
$-(C_6H_4-4-NHNH_2)$	−0.79
$-O-\text{cyclo}-C_6H_{11}$	−0.35
$-O-(CH_2)_5CH_3$	−0.32
$-OCH(CH_3)C(CH_3)_3$	−0.5
C_6; *4 elements*	
$-(C_6H_4-3-NO_2)$	0.10
$-(C_6H_4-4-NO_2)$	0.13
C_6; *5 elements*	
$-(C_6H_4-4-SO_2NH_2)$	0.00
C_7; *2 elements*	
$-CH_2C_6H_5$	−0.69
$-(C_6H_4-3-CH_3)$	−0.55
$-(C_6H_4-4-CH_3)$	−0.60
C_7; *3 elements*	
$-(C_6H_4-4-CN)$	−0.04
$-(C_6H_4-3-COO^-)$	−0.42
$-(C_6H_4-4-COO^-)$	−0.42
$-(C_6H_4-3-COOH)$	−0.18
$-(C_6H_4-4-COOH)$	−0.14
$-(C_6H_3-3,4-OCH_2O-)$	−0.55
$-(C_6H_4-3-OCH_3)$	−0.45
$-(C_6H_4-4-OCH_3)$	−0.59
$-O(C_6H_4-4-CH_3)$	−0.14
$-CH_2OC_6H_5$	−0.2
$-NH(C_6H_4-4-CH_3)$	−1.7
$-(C_6H_4-3-NHCH_3)$	−0.65
$-(C_6H_4-4-NHCH_3)$	−0.81
$-(C_6H_4-3-NH_2CH_3{}^+)$	0.3
C_8; *2 elements*	
$-C\equiv CC_6H_5$	0.28
$-CH=CHC_6H_5$	−0.58
$-CH_2CH_2C_6H_5$	−1.06
$-(CH_2)_7CH_3$	−1.11

Table 8.2 (*Contd.*)

Substituent	σ^{ϕ}
C_8; *3 elements*	
$-(C_6H_4-4-COOCH_3)$	−0.24
$-(C_6H_4-4-SC_2H_5)$	−0.50
$-(C_6H_4-4-OC_2H_5)$	−0.66
$-(C_6H_4-3-NHC_2H_5)$	−0.62
$-[C_6H_4-4-N(CH_3)_2]$	−0.68
$-[C_6H_4-4-N(CH_3)_2H^+]$	0.3
C_8; *4 elements*	
$-(C_6H_4-4-NHCOCH_3)$	−0.49
C_{10}; *3 elements*	
$-(C_6H_4-3-NHC_4H_9)$	−0.67
C_{12}; *2 elements*	
$-(CH_2)_{11}CH_3$	−1.24
C_{13}; *2 elements*	
$-CH(C_6H_5)_2$	−0.73
C_{13}; *3 elements*	
$-(C_6H_4-4-OCOC_6H_5)$	−0.02
C_{19}; *2 elements*	
$-C(C_6H_5)_3$	−1.02

constants of phosphoric, phosphonic, phosphinic, phosphonous and dithioic phosphorus acids show linear correlations with ϕ.

Prediction of pK_a values is straightforward. For example, consider butylphosphonic acid, $C_4H_9(O)P(OH)_2$.

Predicted pK_a for XYP(O)OH $= 1.00 - 0.99\,\sigma^{\phi}$ (Table 8.3)
σ^{ϕ} for X = butyl is -1.22; σ^{ϕ} for Y = $-$OH is -0.39
Predicted p$K_a = 1.00 + 1.61 - 0.3\dagger = 2.31$.

Experimental value is 2.59.

The second pK_a is predicted using the equation:

$$\text{p}K_a \text{ for XP(O)(O}^-\text{)OH} = 6.13 - 1.79\,\sigma^{\phi} \quad \text{(Table 8.3)}$$
$$= 6.13 - 1.79(-1.22) = 8.31.$$

† Statistical factor for proton loss from two identical sites.

Table 8.3 *Hammett equations for substituted phosphorus acids (Mastryukova and Kabachnik, 1969)*

Acids	$pK_a =$
XYPOOH	$1.00 - 0.99\sigma^{\Phi}$
$XP(O)(O^-)(OH)$	$6.13 - 1.79\sigma^{\Phi}$
XP(H)OOH	$1.14 - 1.20\sigma^{\Phi}$
XYPSOH	$0.70 - 0.92\sigma^{\Phi}$, in aqueous 7% ethanol
XYPSeOH	$0.56 - 0.88\sigma^{\Phi}$, in aqueous 7% ethanol
XYPSSH	$1.74 + 0.01\sigma^{\Phi}$, in aqueous 7% ethanol
para-$XYP(O)C_6H_4COOH$	$4.88 - 0.02\sigma^{\Phi}$, in aqueous 50% ethanol†
meta-$XYP(O)C_6H_4COOH$	$5.05 - 0.04\sigma^{\Phi}$, in aqueous 50% ethanol†

† Equation is for −COOH group

[For acids of the form $XP(O)(O^-)OH$, the statistical factor is allowed for in the constant term, 6.13. In the first pK_a, the substituent, Y, was not necessarily −OH, and the constant term, 1.00, did not include the statistical factor].

$$\text{Experimental } pK_a = 8.43$$

Similarly, the pK_a of dibutylphosphinic acid, $(C_4H_9)_2(O)POH$, is predicted to be $1.00 - 0.99\,(2 \times -1.22) = 3.42$. The experimental value is 3.41.

Chapter Nine

Examples where Prediction Presents Difficulties

It is important to recognize limitations to the possibility of predictions and to be on the alert for methods by which these inadequacies can be identified. This is often possible if data on appropriate model compounds are available.

9.1 *Cis* and *trans* isomers

The distance between reaction centres in geometrical isomers varies, depending on whether they are *cis* or *trans* to each other. *Trans* isomers have greater separation and approximate to conditions in saturated systems. Differences between the pK_a values of *cis* and *trans* isomers are enhanced when other substituents are also present. When maleic acid ($pK_a = 2.00, 6.26$) and fumaric acid ($pK_a = 3.01, 5.39$) are compared, the latter approximates more closely to malonic acid ($pK_a = 2.85, 5.70$), in which the transmission of the inductive effect through the $-CH_2-$ is comparable with the effect across $-CH{=}CH-$. Similarly, glutaconic acid ($pK_a = 3.75, 5.08$) compares with succinic acid ($pK_a = 4.22, 5.62$).

9.2 Multiple pK_a values in small molecules

Where there are more than two acidic or basic centres in a small molecule such as citric acid (*9.1*) the predicted and measured pK_a values for the third and higher reaction centres fall off rapidly in their 'goodness of fit'.

```
      H2C—COOH
         |
   HO—C—COOH
         |
      H2C—COOH
```

(*9.1*)

9.3 Steric factors and strain

Although azetidine (*9.2*) and pyrrolidine (*9.3*) have pK_a values of 11.29 and 11.27, the pK_a of aziridine (*9.4*) is only 8.01. This is probably because of the considerable ring-strain in the 3-membered ring which makes it more difficult to alter bond angles when protonation occurs. The weakness of neostrychinine (*9.5*) as a base may be similarly explained.

(*9.2*) 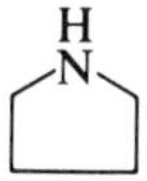 (*9.3*)  (*9.4*) 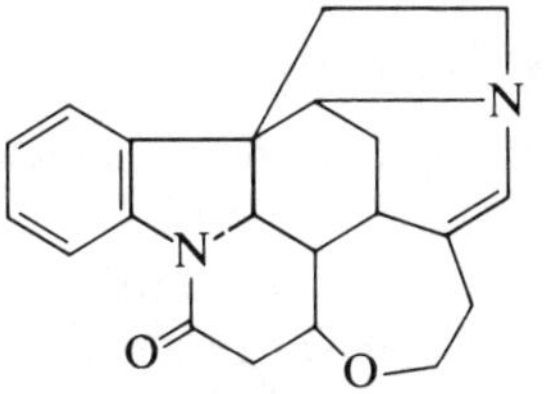 (*9.5*)

9.4 Covalent hydration

In N-heteroaromatic molecules which are very deficient in π-electrons (for example, pteridine), one or more of the $-C=N-$ double bonds can become sufficiently polarized that a water molecule is added covalently across it to give $-\underset{\text{OH}}{\underset{|}{C}}-\underset{\text{H}}{\underset{|}{N}}-$. The covalent hydration in quinazoline cation is represented:

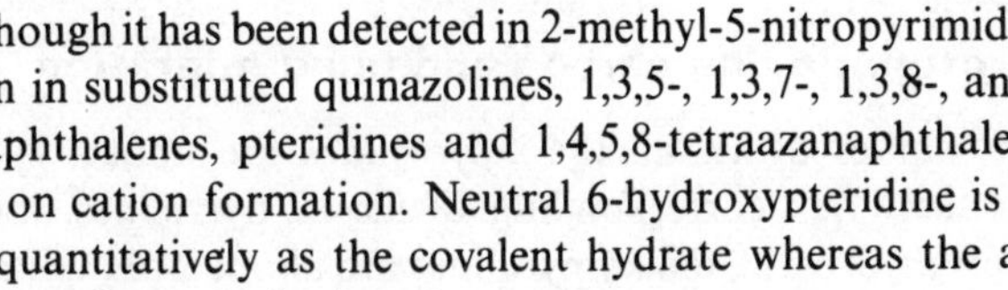

$$+ H^+ + H_2O \rightleftharpoons$$

This reaction is rare in heteroaromatics containing only a single ring, although it has been detected in 2-methyl-5-nitropyrimidine. It is common in substituted quinazolines, 1,3,5-, 1,3,7-, 1,3,8-, and 1,4,6-triazanaphthalenes, pteridines and 1,4,5,8-tetraazanaphthalenes, especially on cation formation. Neutral 6-hydroxypteridine is present almost quantitatively as the covalent hydrate whereas the anion is almost completely 'anhydrous'.

The hydrated species are much stronger bases (2-6 pK_a units) and weaker acids than the anhydrous forms. The ordinarily measured equilibrium pK_a is not readily predictable, but where rapid-reaction studies have made it possible to measure the true pK_a values of such

Table 9.1 *Equilibrium and rapid-reaction* pK_a *measurements for some nitrogen heterocycles*

Base	Predicted pK_a	Apparent pK_a	True pK_a†
Quinazoline	1.3	3.46	1.95
1,3,5-Triazanaphthalene	0.2	4.11	1.36
1,3,7-Triazanaphthalene	1.6	4.70	1.81
1,3,8-Triazanaphthalene	0.4	3.85	~1.0
1,4,6-Triazanaphthalene	1.7	4.60	2.5
Pteridine	−2.5	4.05	–
1,4,5,8-Tetraazanaphthalene	−3.5	2.47	–

† From rapid-reaction measurements (Bunting and Perrin, 1966).

molecules, predicted pK_a values have been in reasonable agreement. Table 9.1 shows results for several heterocyclic bases.

Conditions favouring covalent hydration are:

(a) the presence in the molecule of two or more doubly bonded nitrogen atoms, especially when they are 1,3- to one another in the same ring;

(b) a considerable level of π-electron depletion;

(c) the possibility of resonance stabilization of the structure as a result of covalent hydration.

The phenomenon is usually detected by hysteresis effects during acid–base titrations, anomalous ultraviolet or p.m.r. spectra, and anomalous pK_a values, including an apparently large base-weakening effect produced by an alkyl substituent on the carbon which is involved in the covalent hydration.

For further qualitative and quantitative discussion see Albert and Armarego (1965) and Perrin (1965a).

9.5 Protonation on carbon and bond migration

The pK_a values of 1,2-dialkyl-2-pyrrolines are about 4 pH units greater than the corresponding 2-alkyl-2-pyrrolines. Similarly, 1,4,5,6-tetrahydro-1,2,-dialkylpyridines are about 2 pH units greater than the corresponding 1-alkyl derivatives. It is suggested that protonation takes place on *carbon-3*, followed by migration of the double bond to 1,2-position, so that the positive charge resides on the nitrogen (Adams and Mahan, 1942). Pyrrole has an α-carbon atom as the preferred site for proton addition, but if this is already alkylated the other α-carbon atom is protonated. A methyl in a β-position favours the adjacent α-site (Chiang and Whipple, 1963). Similar considerations apply to the protonation of indoles, indolizines and other π-electron-excessive N-heteroaromatic compounds.

9.6 Keto–enol tautomerism

Substituent effects of $-OH$ and $-OCH_3$ groups are usually similar to each other. Nevertheless they differ considerably in the pyridine series (as shown by their σ_{ortho} and σ_{para} constants), due to a strong preference for the α- and γ-hydroxy derivatives to form cyclic ketones in which the proton is on the nitrogen instead of the oxygen atom. They are very weak bases (pK_a 1– –2) and weak acids. The differences are even greater in the sulphur analogues.

Similarly, imines can be formed by α- and γ-heterocyclic amines, but the weaker amines are the preferred species. N-alkylation requires formation of the imino structure and the resulting base is comparable in strength with amidine analogues. O-methylation of urea is similarly base-strengthening.

9.7 The principle of Vinylogy (Fuson, 1935)

Because 2-aminopyridine is a cyclic amidine, its cation is stabilized by 'amidinium'-type electron delocation, making this amine a much stronger base than pyridine or 3-aminopyridine:

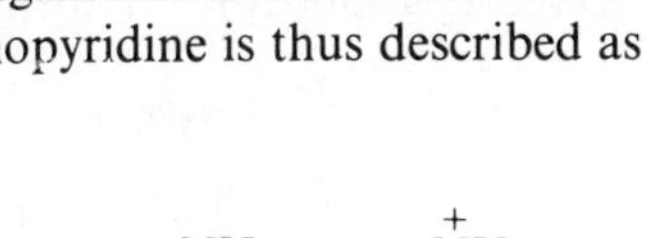

In 4-aminopyridine, the conjugation pathway is extended by a vinyl group which once again makes for increased electron delocalization in the cation. 4-Aminopyridine is thus described as a vinylogous cyclic amidine:

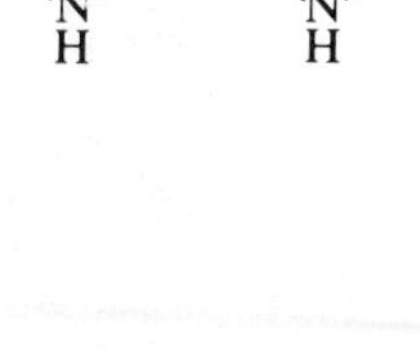

4-Aminopyridine (pK_a = 9.11) is a stronger base than 2-aminopyridine (pK_a = 6.71).

Similarly, 1-anilino-5,5-dimethyl-3-phenyliminocyclohex-1-ene (*9.6*) is a vinylogous amidine and hence is a stronger base (pK_a = 9.89) than aniline (pK_a = 4.80). The conjugation pathway may contain more double bonds, as for example, in 1-methylamino-7-methyliminocyclohepta-1,3-5-triene (*9.7*)(pK_a = 10.2), and this also increases the strength of the base.

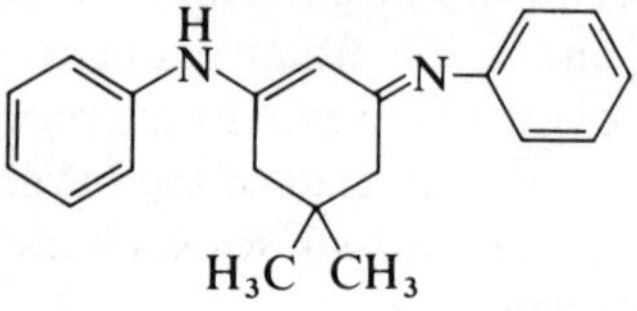

(*9.6*) 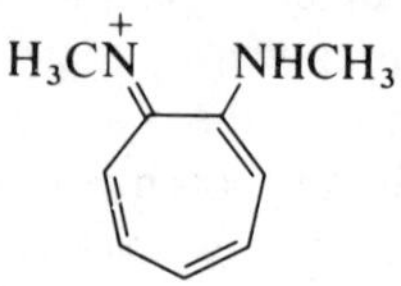 (*9.7*)

3-Hydroxy-2,4-dimethyl-2-cyclobutenone is a vinylogous carboxylic acid, forming the anion:

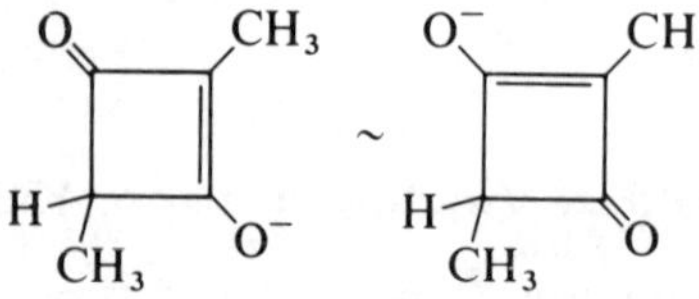

Its increased conjugation pathway makes this acid (pK_a = 2.8) about 100 times stronger than the corresponding carboxylic acid (cf. acetic acid, pK_a = 4.76). Other examples include squaric acid (*9.8*) (pK_a = 1.5, 3.48), 4,5-dihydroxy-4-cyclopentene-1,2,3-trione (*9.9*) (pK_a = 0.89, 3.06), 3,5,6-tribromo-2-hydroxy-1,4-benzoquinone (*9.10*)
(pK_a = 1.10), 3-acetyl-4-hydroxy-6-methyl-2*H*-pyran-2-one (*9.11*) (pK_a = 5.12) and ascorbic acid(*9.12*) (pK_a = 4.52). Tetronic acid (*9.13*) (pK_a = 3.76) serves as a model for ascorbic acid.

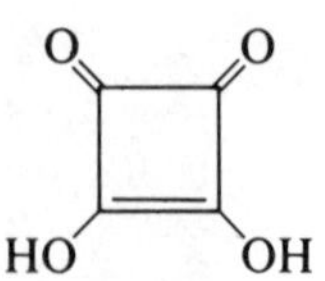

(*9.8*) 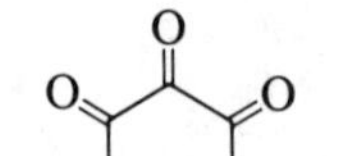 (*9.9*) 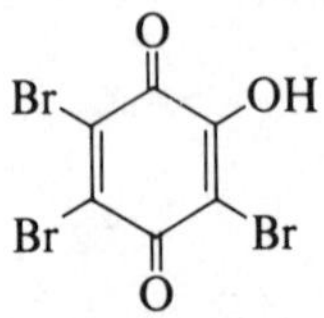 (*9.10*)

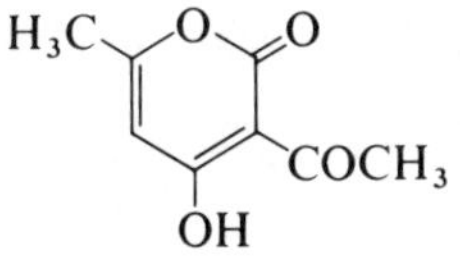

(*9.11*) 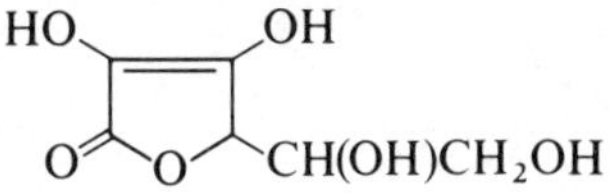 (*9.12*) 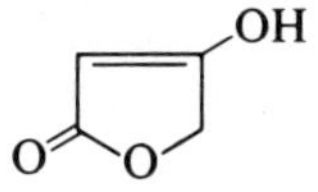 (*9.13*)

The enhanced acidity when a hydroxyl group is *meta* to a carbonyl function in a nitrogen heterocycle is shown by 1,2,3,6-tetrahydro-2,6-dioxo-4-hydroxypyrimidine (*9.14*) ($pK_a = 3.99$). The long conjugation pathway in tropolone (*9.15*) ($pK_a = 6.69$) stabilizes the anion. The same effect is shown by the 2-mercapto derivative ($pK_a = 5.90$).

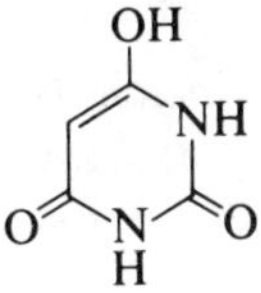

(*9.14*) 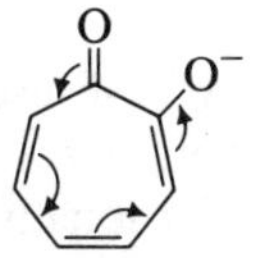 (*9.15*)

An aldehydic carbonyl group can be involved, for example, in 2-hydroxy-3,5-dinitrobenzaldehyde (*9.16*) ($pK_a = 2.27$) and 3,5-dichloro-4-hydroxybenzaldehyde (*9.17*) ($pK_a = 4.25$).

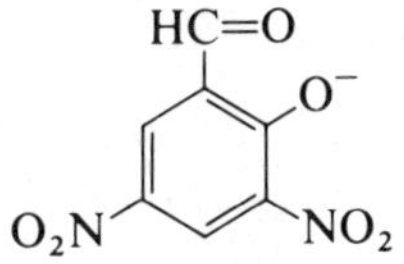

(*9.16*) 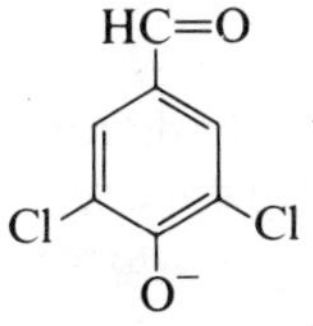 (*9.17*)

These observations can be generalized as follows: if one or more component parts of an acidic or basic group are separated by one or more vinyl groups, the acidic or basic group may still retain its activity, wholly or in part, and may, if the electron delocalization is great enough, even show enhanced activity.

The weakness of amides as bases is much less apparent in their vinylogues. Thus, although the pK_a of 1-methyl-1,2-dihydro-2-oxopyridine (*9.18*) is 0.32, the pK_a of the vinylogous amide, 1-methyl-1,4-dihydro-4-oxopyridine (*9.19*), is 3.29 (corresponding to a thousand-fold increase in strength as a base).

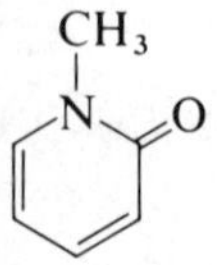

(*9.18*) 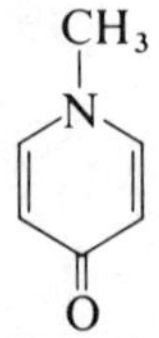 (*9.19*)

The explanation probably lies in a much smaller contribution by 'resonance' to the stabilization of the amidium cation, with the charge distributed unequally over the nitrogen and oxygen atoms, so that the main effect observed is that with increasing distance along the carbon chain, the pK_a values converge towards those expected for the parent amines.

As an example of the use of the principle of vinylogy in pK_a prediction, consider tetracycline which, for this purpose, can be divided into three portions as shown in (*9.20*). Each portion contains

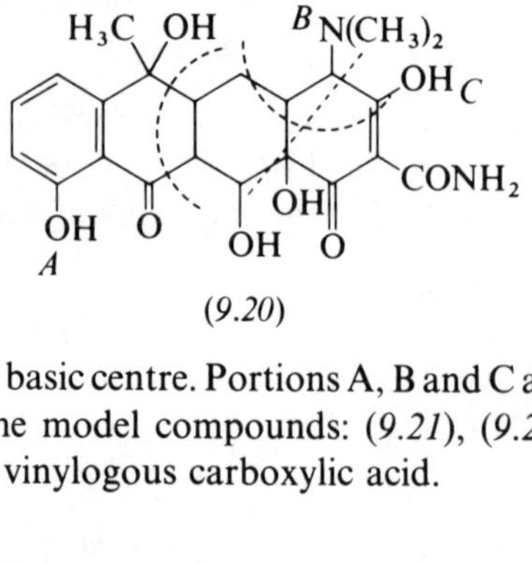

(*9.20*)

only one acidic or basic centre. Portions A, B and C are approximated respectively by the model compounds: (*9.21*), (*9.22*) and (*9.23*) of which (*9.23*) is a vinylogous carboxylic acid.

$CH_3COC(CONH_2){=}CHOH$

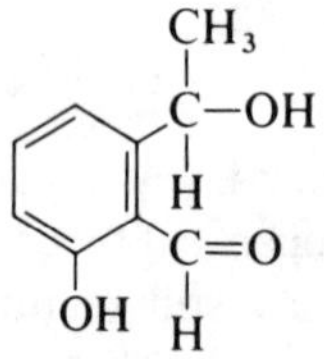

(*9.21*) 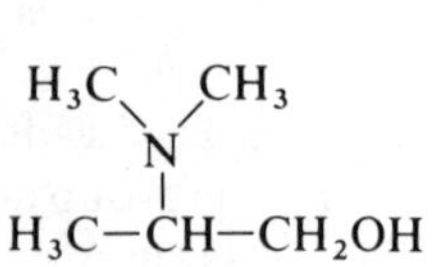 (*9.22*) 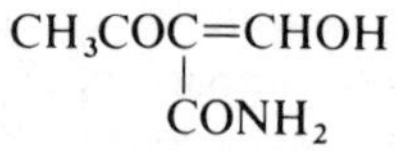 (*9.23*)

It would be reasonable to expect that the pK_a of the acid (*9.23*) would lie in the range of values for the vinylogous carboxylic acids previously quoted, i.e. 0.89–5.12. However, it is structurally more similar to (*9.11*) and so its pK_a would be expected to lie in the upper part of the range, say 4–5.

The pK_a of the phenol (*9.21*) is predicted as 8.07 using $pK_a = 9.92 - 2.23\,(\sigma_{ortho}$ for $-CHO + \sigma_{meta}$ for $-CH(OH)CH_3)$. For the tertiary amine (*9.22*), the use of Equation (3.4) and $-\Delta pK_a$ for the β-OH (Table 4.2) gives a predicted pK_a of 9.0.

The experimental values for tetracycline are:

3.3 (cf. ~ 4–5 predicted on the basis of being a vinylogous carboxylic acid similar to (*9.11*);

7.76 (cf. 8.07 predicted for phenolic –OH);

and

9.64 (cf. 9.0 predicted for tertiary amine group).

Chapter Ten

Recapitulation of the Main pK_a Prediction Methods

When a pK_a value is required, it may be possible to find the value at the required temperature and ionic strength in one of the compilations listed in Chapter 1. If the required value is not available, the ideal procedure would be to determine the value under the required conditions using procedures described by Albert and Serjeant (1971). If time does not permit or when a less accurately known value will suffice, a prediction based on methods described in this work can be made. It should be noted that the best predictions are those based on linear free energy relationships. In general, predictions are best which use the least number of substituent sigma constants and the least number of approximations. Ideally, pK_a predictions should be based on a model compound as similar as possible in structure to the required compound.

The preceding chapters set out, in some detail, methods by which to predict pK_a values. The main features of these methods are summarized in the following sections.

10.1 Prediction using Hammett and Taft data

Firstly identify the acidic and basic centres (Table 1.2 gives individual pK_a values for representative acids and bases and Table 1.1 gives typical ranges of pK_a values for classes of acid and base).

Then choose the appropriate Hammett or Taft equation from Tables A.2, A.3, and, for phosphorus acids, Table 8.3. Where a choice of equation is available, choose the one which involves the least number of substituent constants. For example, to predict the pK_a of 2-chloro-5-methoxy-4-methylthiopyridinium ion, the equation: $pK_a = 5.54 - 7.33\,\sigma_{ortho}$ (Besso *et al.*, 1977) would be used in preference to the equation: $pK_a = 5.25 - 5.90\,\Sigma\sigma$ (Clark and Perrin, 1964).

The main compilations of sigma substituent constants for use in Hammett and Taft equations are given in Tables A.1, A.4 and A.5, and for phosphorus acids, Table 8.2. The method of treating annelated rings as substituents is discussed in Chapter 7. Table 7.1 gives the appropriate constants when the annelated ring is unsubstituted while Table 7.2 gives constants for naphthalene derivatives when the annelated ring is substituted.

Heteroatoms can be regarded as substituents as discussed in Chapter 8. Table A.6 gives appropriate substituent constants.

When a Hammett or Taft equation is not available for a particular class of acid or base, it may be possible to construct such an equation using one of the methods discussed in the next section.

10.2 Constructing a Hammett or Taft equation

When the pK_a values of related compounds are available, it may be possible to construct a linear free energy relation, pK_a versus appropriate sigma constants, from which to read off the pK_a value that is sought. Suppose that an estimate of the pK_a of the weakly acid ring –NH– in 5-aminotetrazole (*10.1*) is required (Table A.2 does not contain an appropriate equation). The substituent is α- to the acidic centre and σ_{ortho} constants for pyridines would appear the best choice of substituent constant.

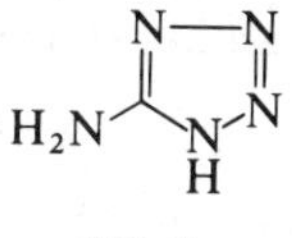

(*10.1*)

The pK_a values of some 5-substituted tetrazoles are known, and the appropriate σ_{ortho} constants are given in Table A.5. These are plotted in Fig. 10.1 where they are seen to fit a line

$$\mathrm{p}K_a = 5.02 - 3.8\,\sigma_{ortho} \qquad (10.1)$$

From the value of σ_{ortho} for $-NH_2$ = −0.27 (Table A.5), the predicted pK_a is read off as 6.1 (the experimental value is 6.00).

Alternatively, a Hammett or Taft equation for a class of compounds can sometimes be developed from a single pK_a value by predicting the ρ- value and then determining the constant term, pK_a^0, by extrapolation as described in Section 8.2.

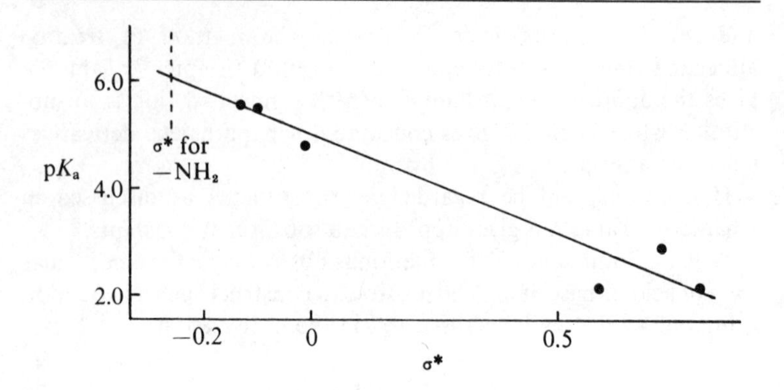

Fig. 10.1 pK_a values of 5-substituted tetrazoles. Least-squares equation: $pK_a = 5.02 - 3.80\ \sigma_{ortho}$

10.3 Where sigma values are not available

Where a required substituent is not available in the extensive list of σ_{meta}, σ_{para} and σ^* constants given in Table A.1, it may be possible to make an estimate using one of the procedures described in earlier chapters. For convenience, these are summarized here.

For example, σ^* values can be estimated from σ_{meta} values, and vice versa, using the empirical relation (Equation (5.3)):

$$\sigma_{meta} = 0.217\ \sigma^* - 0.106$$

When σ^* for a substituent $-R$ is known, σ^* for $-CH_2R$ can be estimated using Equation (4.5):

$$\sigma^* \text{ for } -CH_2R \sim 0.4 \times \sigma^* \text{ for } -R$$

An extension of this relation:

$$\sigma^* \text{ for } -(CH_2)_nR \sim (0.4)^n \times \sigma^* \text{ for } -R$$

has been used in several worked examples including Sections 4.4.2 and 4.4.4.

Another approximation of use is given in Equation (4.6):

$$\sigma^* \text{ for } -CXYZ \sim \sigma^* \text{ for } -CH_2X + \sigma^* \text{ for } -CH_2Y + \sigma^* \text{ for } -CH_2Z$$

For specific applications, see worked examples in Sections 4.4.7 and 8.2.2.

Another procedure is to choose a sigma constant for a substituent which is structurally similar to the required group. In the worked example of Section 6.4.1, the group, $-OCH(CH_2OH)$-$CHOHCHOH-$ is approximated by $-OCH(CH_3)CH_2CH_3$ for which σ^* is available in Table A.1.

A method for estimating the σ^* value of an (X-substituted) phenyl from the appropriate sigma constant for $-X$ is illustrated in the following example:

$$\sigma_{meta} \text{ for } -NO_2 = 0.74 \qquad \text{(Table A.1)}$$

Predicted pK_a for
3-nitrophenylacetic acid $= 4.30 - 0.49(\sigma_{meta}$ for $-NO_2)$ (Table A.2)
$= 3.94$

Predicted pK_a for an acetic acid,
$RCH_2COOH = 4.76 - 0.67\sigma^*$ (Table A.2)
$= 3.94$

Therefore, σ^* for 3-nitrophenyl $= (4.76 - 3.94)/0.67 = 1.22$

The estimates are least satisfactory when X- is a charged substituent.

10.4 The ΔpK_a method

The ΔpK_a method is a useful alternative to the Taft equation approach for aliphatic carboxylic acids and aliphatic amines. In addition, the approach can be used with pyridines. The method is described in Chapter 4 and Tables 4.1, 4.2 and 4.3 give appropriate constants. Numerous worked examples are given in Section 4.2.

If ΔpK_a values are not available in Table 4.1 (aliphatic acids) they may be calculated using Equation (4.1):

$$-\Delta pK_a = 0.06 + 0.63\sigma^*$$

where σ^* values are obtained from Table A.1, and noting that the value obtained applies to the α-position. Values for aliphatic amines, when not available in Table 4.2, may be calculated using Equation (4.2):

$$-\Delta pK_a = 0.28 + 0.87\sigma^*$$

in this case noting that the value is for the β-position.

The method can be extended for substituents of the form, $R(CH_2)_n-$, since each $-CH_2-$ attenuates the effect of a substituent, $R-$, by a factor of approximately 0.4. See Equation (4.3).

10.5 Analogy, modelling and fragmentation

Section 3.3 discusses pK_a prediction by reference to compounds with molecular structures which are different from, but related to, the particular compounds of interest. Further examples are given in Section 7.1. See also the worked example of Section 4.2.10.

For complex molecules, it is often convenient to make a notional division of the structure into molecular fragments which are amenable to prediction using methods summarized in Sections 10.1–10.4. Several examples are discussed in detail in Section 7.4.

10.6 Miscellaneous relationships

A number of useful relations described in earlier chapters are gathered in the following list:

Equation (5.7): $(\mathrm{p}K_a)^{\mathrm{subst}}_{\mathrm{phenol}} = 0.72\,(\mathrm{p}K_a)^{\mathrm{subst}}_{\mathrm{aniline}} + 6.46$

Equation (6.1): $(\mathrm{p}K_a)^{\mathrm{subst}}_{\mathrm{quinoline}} = 2.1\,(\mathrm{p}K_a)^{\mathrm{subst}}_{\alpha\text{-naphthylamine}} - 3.1$

Equation (6.2): $(\mathrm{p}K_a)^{\mathrm{subst}}_{\mathrm{isoquinoline}} = 2.1\,(\mathrm{p}K_a)^{\mathrm{subst}}_{\beta\text{-naphthylamine}} - 3.7$

10.7 Polybasic acids

The ability to estimate pK_a values with even a reasonable degree of accuracy is especially useful in considering the sites and sequences of protonation and deprotonation with changing pH. The general principle is described and specific examples considered in Section 7.3. Worked examples in Sections 4.2.5, 5.3.5, 5.3.6, 6.2.6 and 6.4.2 are also relevant. Similar issues apply in predicting the presence, or absence, of zwitterionic species. The conditions for zwitterion formation are also described in Section 7.3, together with a detailed study of two examples. See also worked example in Section 4.2.9.

When an acid (or base) has two or more structurally equivalent acidic (or basic) centres, a statistical effect must be taken into account as discussed in Section 2.4. Worked examples in which statistical effects are involved include Sections 4.2.7, 4.2.10, 4.4.4, 5.3.2, 5.3.3 and 6.2.4.

10.8 Ionic strength and temperature corrections

If a pK_a is required at a temperature other than 25° and at a given ionic strength, the corrections described in Sections 1.5 and 1.4, respectively, are applied.

Appendix

Table A.1 comprises an extensive compilation of Hammett σ and Taft σ^* values, while Tables A.2 and A.3 give numerous Hammett and Taft equations for organic acids and bases. Taken in conjunction they enable pK_a values for a wide range of organic acids and bases to be predicted.

Table A.4 lists σ constants that apply to anilines, phenols and pyridines in cases where there is significant electron delocalization resulting from the *para* substituent. Extension to *ortho* substituted acids and bases is possible by using the 'apparent' σ_{ortho} constants given in Table A.5. Sigma constants for heteroatoms in heterocyclic rings are listed in Table A.6.

A.1 Substituent constants for the Hammett and Taft equations (Table A.1)

The substituents in the table are arranged in groups on the basis of the number of carbon atoms and on the number of elements involved. Within each group, the sequence of substituents is based on the alphabetical order of the element symbols except that, in carbon containing substituents, hydrogen is placed immediately after carbon. The molecular formula of each substituent is given together with a condensed structural formula (or, in a few cases, a name). Within groups of isomeric substituents, the sequence is arbitrary but is generally based on systematic names. At the end of the Table, data for some generalized substituent formulae (e.g. – OR) are given under the heading C_n, for use when the substituent of interest is not otherwise available.

Substituent		σ_{meta}	σ_{para}	σ^*
C_0; *1 element*				
Br	– Br	0.39	0.22	2.84
Cl	– Cl	0.37	0.24	2.96
F	– F	0.34	0.06	3.21

Table A.1 (*Contd.*)

Substituent		σ_{meta}	σ_{para}	σ^*
C_0; *1 element* (*Contd.*)				
H	–H	0.00	0.00	0.49
I	–I	0.35	0.21	2.46
N_3	$-N_3$	0.27	0.15	2.62
O	$-O^-$	–0.47	–0.81	–2.78
S	$-S^-$	–0.36		
C_0; *2 elements*				
BF_2	$-BF_2$	0.32	0.48	
Br_3Ge	$-GeBr_3$	0.66	0.73	3.7
Br_3Si	$-SiBr_3$	0.48	0.57	2.4
ClHg	–HgCl	0.33	0.35	1.9
ClO_3	$-ClO_3$	0.85	†	
ClS	–SCl	0.44	0.48	2.50
Cl_2I	$-ICl_2$	1.10	1.11	6.8
Cl_2P	$-PCl_2$	0.53	0.61	2.84
Cl_3Ge	$-GeCl_3$	0.71	0.79	3.9
Cl_3Si	$-SiCl_3$	0.48	0.56	2.4
FHg	–HgF	0.34	0.33	2.1
F_2I	$-IF_2$	0.85	0.83	5.4
F_2P	$-PF_2$	0.48	0.59	2.37
F_3Ge	$-GeF_3$	0.85	0.97	4.6
F_3S	$-SF_3$	0.70	0.80	3.75
F_3Si	$-SiF_3$	0.54	0.66	2.62
F_4I	$-IF_4$	1.07	1.15	6.3
F_4P	$-PF_4$	0.63	0.80	2.8
F_5S	$-SF_5$	0.61	0.68	3.56
GeH_3	$-GeH_3$	0.00	0.01	~0.7
HO	–OH	0.13	–0.38†,‡	1.34
HS	–SH	0.25	0.15§	1.68
H_2N	$-NH_2$	0.00	–0.57†	0.62
H_2P	$-PH_2$	0.06	0.05	
H_3N	$-NH_3^+$	0.67	0.53	3.76
H_3N_2	$-NHNH_2$	–0.02	–0.55	0.40
H_3Si	$-SiH_3$	0.05	0.10	
IO_2	$-IO_2$	0.70	0.76	3.71
NO	–NO			2.08
NO_2	$-NO_2$	0.74	0.78	4.25
NO_3	$-ONO_2$	0.55	0.70	3.86
O_2S	$-SO_2^-$	–0.02	–0.05	
O_3P	$-PO_3^{2-}$	–0.02		
O_3S	$-SO_3^-$	0.31	0.37†,‡§	0.81
C_0; *3 elements*				
$AsHO_3$	$-AsO_3H^-$	0.00	–0.02	0.14
BH_2O_2	$-B(OH)_2$	–0.01	0.12	0.95
ClOS	–SOCl	0.75	0.82	4.25
ClO_2S	$-SO_2Cl$	0.92	1.04	5.0
Cl_2OP	$-POCl_2$	0.78	0.90	4.1

Table A.1 (*Contd.*)

Substituent		σ_{meta}	σ_{para}	σ^*
C_0; *3 elements* (*Contd.*)				
Cl_2PS	$-PSCl_2$	0.70	0.80	3.7
FOS	$-SOF$	0.74	0.83	4.12
FO_2S	$-SO_2F$	~0.89	~0.99	~4.7
F_2OP	$-POF_2$	0.81	0.89	
F_5OS	$-OSF_5$		0.44	
HO_2P	$-PO(OH)^-$		0.14	
HO_2S	$-SO(OH)$	−0.04	−0.07	
HO_3P	$-PO_3H^-$	0.25	0.17	1.41
HO_3S	$-SO_2OH$	0.55		
HO_4P	$-OPO_3H^-$	0.29	0.00	
H_2NO	$-NHOH$	−0.04	−0.34	0.30
H_2O_3P	$-PO(OH)_2$	0.36	0.42	
H_3NO	$-ONH_3^+$	0.53		2.92
C_0; *4 elements*				
H_2NO_2S	$-SO_2NH_2$	0.46	0.57‡	2.61
C_1; *2 elements*				
CBr_3	$-CBr_3$	0.28	0.29	2.43
CCl_3	$-CCl_3$	0.40	0.46	2.65
CF_3	$-CF_3$	0.46	0.53‡	2.61
CH_3	$-CH_3$	−0.06	−0.14	0.00
CN	$-CN$	0.62	0.70†,‡,§	3.30
	$-NC$	0.48	0.49	
CN_7	5-azido-1-tetrazolyl	0.54	0.54	
CO_2	$-COO^-$	0.09	−0.05	−1.06
C_1; *3 elements*				
$CClF_2$	$-CClF_2$	0.42	0.46	2.37
$CClN_4$	5-Cl-1-tetrazolyl	0.80	0.61	
CClO	$-COCl$	0.53	0.69	2.37
CCl_2N	$-N=CCl_2$	0.21	0.13	1.81
CCl_3O	$-OCCl_3$	0.43	0.35	3.19
CFO	$-COF$	0.55	0.70	2.44
CF_2O_2	3, 4-OCF_2O-	—(0.36)—		
CF_3Hg	$-HgCF_3$	0.29	0.32	1.7
CF_3N_2	$-N=NCF_3$	0.56	0.68	2.75
CF_3O	$-OCF_3$	0.36	0.33	
CF_3S	$-SCF_3$	0.38	0.50	2.75
CF_3Se	$-SeCF_3$	0.32	0.38	2.62
$CHBr_2$	$-CHBr_2$	0.31	0.32	1.96
$CHCl_2$	$-CHCl_2$	0.31	0.32	1.94
CHF_2	$-CHF_2$	0.32	0.35	2.05
CHI_2	$-CHI_2$	0.26	0.26	1.62
CHN_2	$-NHCN$	0.21	0.06	
CHN_4	1-tetrazolyl	0.60	0.57	3.4
CHO	$-CHO$	0.36	0.44†,‡,§	2.15
CHO_2	$-COOH$	0.35	0.44	2.08

Table A.1 (*Contd.*)

Substituent		σ_{meta}	σ_{para}	σ^*
C_1; *3 elements* (*Contd.*)				
CH_2Br	$-CH_2Br$	0.12	0.14	1.00
CH_2Cl	$-CH_2Cl$	0.11	0.12	0.94
CH_2F	$-CH_2F$	0.11	0.10	1.10
CH_2I	$-CH_2I$	0.10	0.11	1.00
CH_2O_2	3,4-OCH_2O-	—(−0.27)—		
CH_3Hg	$-HgCH_3$	0.43	0.10	
CH_3O	$-CH_2OH$	0.01	0.01	0.62
	$-OCH_3$	0.11	−0.28†,§	1.81
CH_3O_2	$-CH(OH)_2$			1.37
CH_3S	$-CH_2SH$	0.08		0.62
	$-SCH_3$	0.14	0.00†,§	1.56
CH_3S_2	$-SSCH_3$	0.22	0.13	
CH_3Se	$-SeCH_3$	0.1	0.00	0.95
CH_4Cl	$-CHClCH_3$			1.00
CH_4N	$-CH_2NH_2$	−0.03	−0.11	0.50
	$-NHCH_3$	−0.30	−0.84	−0.81
CH_5N	$-CH_2NH_3^+$	0.32	0.29	2.24
	$-NH_2CH_3^+$	0.96		3.74
CHgN	−HgCN	0.28	0.34	1.4
CNO	−OCN	0.67	0.54	5.0
	−NCO	0.27	0.19	2.25
CNS	−NCS	0.48	0.38	2.62
	−SCN	0.41	0.52	3.43
CNSe	−SeCN	0.61	0.66	3.61
CN_3O_6	$-C(NO_2)_3$	0.72	0.82	4.6
C_1; *4 elements*				
CF_3HgS	$-HgSCF_3$	0.39	0.42	2.3
CF_3OS	$-S(O)CF_3$	0.63	0.69	4.30
CF_3OSe	$-SeOCF_3$	0.81	0.83	4.75
CF_3O_2S	$-SO_2CF_3$	0.76	0.95	4.50
CF_3O_2Se	$-SeO_2CF_3$	1.08	1.21	6.0
CF_3O_3S	$-OSO_2CF_3$	0.56	0.53	4.37
$CHCl_2O$	$-OCHCl_2$	0.38	0.26	3.06
CHF_2O	$-OCHF_2$	0.31	0.18	2.81
CHF_2S	$-SCHF_2$	0.33	0.36	2.06
CHN_4O	5-OH-1-tetrazolyl	0.39	0.33	
CHN_4S	5-SH-1-tetrazolyl	0.45	0.45	
	1,2,3,4-Thiatriazol-5-ylamino	0.30	0.19	2.62
CH_2ClO	$-OCH_2Cl$	0.25	0.08	2.56
CH_2FO	$-OCH_2F$	0.20	0.02	2.31
CH_2FS	$-SCH_2F$	0.23	0.20	1.69
CH_2NO	$-CONH_2$	0.28	0.31	1.68
	−NHCHO	0.19	0.00	1.62
	−CH = NOH	0.22	0.10	
CH_3Br_2Si	$-SiBr_2(CH_3)$		0.29	
CH_3Cl_2Si	$-SiCl_2(CH_3)$	0.31	0.39	1.5
CH_3F_2Si	$-SiF_2(CH_3)$	0.29	0.23	

Table A.1 (*Contd.*)

Substituent		σ_{meta}	σ_{para}	σ^*
C_1; *4 elements* (*Contd.*)				
CH_3N_2O	$-NHCONH_2$	−0.03	−0.24	1.31
CH_3N_2S	$-NHCSNH_2$	0.22	0.16	1.8
CH_3OS	$-SOCH_3$	0.21	0.17	1.56
	$-S(O)CH_3$	0.52	0.49†	2.88
CH_3O_2S	$-SO_2CH_3$	0.64	0.73†,‡	3.68
	$-S(O)OCH_3$	0.50	0.54	2.84
	$-OS(O)CH_3$	0.44	0.45	
$CH_3O_2S_2$	$-SSO_2CH_3$	0.43	0.54	
CH_3O_3S	$-OSO_2CH_3$	0.39	0.36	3.62
CNO_2S	$-SO_2CN$	1.10	1.26	5.9
C_1; *5 elements*				
CHF_2OS	$-SO(CHF_2)$	~0.70	~0.76	4.1
CHF_2O_2S	$-SO_2(CHF_2)$	0.75	0.86	3.69
CH_2FO_2S	$-SO_2CH_2F$			3.44
CH_2NOS	$-SCONH_2$	0.34		2.07
CH_4NOS	$-NHSO_2CH_3$	0.20	0.03	
C_1; *6 elements*				
CHF_3NO_2S	$-NHSO_2CF_3$	0.44	0.39	3.1
C_2; *2 elements*				
C_2Cl_3	$-CCl=CCl_2$			2.25
C_2F_3	$-CF=CF_2$			1.94
C_2F_5	$-CF_2CF_3$	0.50	0.52	2.56
C_2H	$-C\equiv CH$	0.20	0.23	2.18
C_2H_3	$-CH=CH_2$	0.08	−0.08	0.56
C_2H_5	$-C_2H_5$	−0.07	−0.15	−0.10
C_2; *3 elements*				
C_2F_3O	$-COCF_3$	0.63	0.80	3.7
$C_2F_3O_2$	$-OCOCF_3$	0.56	0.46	4.06
C_2F_4O	$3,4-CF_2OCF_2-$	—(0.81)—		
C_2F_5O	$-OCF_2CF_3$	0.48	0.28	
C_2F_6N	$-N(CF_3)_2$	0.47	0.53	3.1
C_2F_6P	$-P(CF_3)_2$	0.60	0.59	3.12
C_2HCl_2	$-CH=CCl_2$	0.11		1.00
C_2HF_2	$-CH=CF_2$			1.19
C_2HS	$-SC\equiv CH$	0.26	0.19	2.00
C_2H_2Cl	$-CH=CHCl$			0.87
$C_2H_2Cl_3$	$-CH_2CCl_3$	0.06		0.75
C_2H_2F	$-CF=CH_2$			1.56
$C_2H_2F_3$	$-CH_2CF_3$	0.16	0.14	0.87
C_2H_2N	$-CH_2CN$	0.16	0.18	1.30
$C_2H_2O_2$	$-CH_2COO^-$			−0.06
C_2H_3O	$-COCH_3$	0.36	0.47†,‡,§	1.81
	$-CH_2CHO$			0.62
	2-oxacyclopropyl	0.05	0.03	

Table A.1 (*Contd.*)

Substituent		σ_{meta}	σ_{para}	σ^*
C_2; *3 elements* (*Contd.*)				
$C_2H_3O_2$	$-OCOCH_3$	0.39	0.31	2.56
	$-CH_2COOH$		−0.07	1.08
	$-COOCH_3$	0.32	0.39‡	2.00
$C_2H_3O_3$	$-OCH_2COOH$		−0.18	
C_2H_3S	2-thiacyclopropyl	0.04	0.01	
	$-SCH=CH_2$	0.18	0.14	1.31
$C_2H_3S_2$	$-SC(S)CH_3$			3.00
C_2H_4Br	$-CHBrCH_3$			1.25
C_2H_4N	1-azacyclopropyl	−0.07	−0.22	
	2-azacyclopropyl	−0.06	−0.10	
$C_2H_4O_2$	$3,4-O(CH_2)_2O-$	—(−0.12)—		
C_2H_5O	$-OC_2H_5$	0.10	−0.24	1.68
	$-CH(OH)CH_3$	0.08	−0.07	0.12
	$-CH_2OCH_3$	−0.10	0.03	0.66
C_2H_5S	$-SC_2H_5$	0.23	0.03	1.56
C_2H_6N	$-N(CH_3)_2$	−0.15	−0.83‡	0.32
	$-NHC_2H_5$	−0.24	−0.61	−0.62
C_2H_6P	$-P(CH_3)_2$	0.05	0.03	
C_2H_6S	$-S(CH_3)_2^+$	1.00	0.90	5.09
C_2H_7N	$-CH_2CH_2NH_3^+$	0.23	0.17	
	$-NH(CH_3)_2^+$	0.84		4.36
	$-NH_2C_2H_5^+$	0.96		3.74
C_2H_7Si	$-SiH(CH_3)_2$	0.01	0.04	
$C_2H_{11}B_{10}$	$1-C_2HB_{10}H_{10}(1,2)$	0.48	0.45	2.4
	$1-C_2HB_{10}H_{10}(1,7)$	0.25	0.33	1.2
	$3-B_{10}H_9C_2H_2(1,2)$	0.20	0.19	
C_2NO	$-COCN$			3.4
C_2N_2P	$-P(CN)_2$	0.82	0.90	4.6
C_2; *4 elements*				
$C_2F_2NO_2$	$-N(COF)_2$	0.58	0.57	3.6
$C_2F_3HgO_2$	$-HgOCOCF_3$	0.50	0.52	3.0
C_2F_3OS	$-SCOCF_3$	0.48	0.46	3.19
C_2F_4NO	$-N(CF_3)(COF)$	0.56	0.56	3.5
C_2HF_4O	$-OCF_2CHF_2$	0.34	0.25	
C_2HF_4S	$-SCF_2CHF_2$	0.38	0.47	
$C_2H_2F_3S$	$-CH_2SCF_3$	0.12	0.15	0.75
C_2H_2NO	$-CH_2NCO$			0.81
$C_2H_2NO_2$	$-CH=CHNO_2$ (*trans*)	0.32	0.26	1.75
C_2H_2NS	$-CH_2SCN$	0.12	0.14	
	$-CH_2NCS$			0.94
$C_2H_2N_3O_6$	$-CH_2C(NO_2)_3$			1.62
$C_2H_3F_3N$	$-NHCH_2CF_3$			1.8
$C_2H_3HgO_2$	$-HgOCOCH_3$	0.39	0.40	2.4
$C_2H_3N_2O_4$	$-C(CH_3)(NO_2)_2$	0.54	0.61	
C_2H_3OS	$-SCOCH_3$	0.39	0.44	2.29
$C_2H_3O_2S$	$-SO_2CH=CH_2$			3.56

Table A.1 (*Contd.*)

Substituent		σ_{meta}	σ_{para}	σ^*
C_2; *4 elements* (*Contd.*)				
C_2H_4NO	$-NHCOCH_3$	0.12	−0.09	1.40
	$-CH_2CONH_2$	0.06	0.07	0.31
	$-CH=NOCH_3$	0.37	0.30	
	2-methyl-2, 3-azaoxacyclopropyl	0.09	0.12	
	$-CONHCH_3$	0.35	0.36	
$C_2H_4NO_2$	$-CH_2CH_2NO_2$	0.00		0.50
$C_2H_4NO_3$	$-CH_2CH_2ONO_2$			0.49
C_2H_4NS	$-CSNHCH_3$	0.30	0.34	
	$-NHCSCH_3$	0.24	0.12	
$C_2H_4NS_2$	$-NHC(S)SCH_3$			2.6
$C_2H_4N_3S$	$-CH=NNHC(S)NH_2$	0.45	0.40	
$C_2H_5N_4S$	$-CH=NNHCONHNH_2$	0.22	0.16	
$C_2H_5O_2S$	$-SO_2C_2H_5$	0.66	0.77	3.74
	$-CH_2SO_2CH_3$	0.18		1.32
C_2H_6BrSi	$-SiBr(CH_3)_2$		0.10	
C_2H_6ClSi	$-SiCl(CH_3)_2$	0.16	0.21	
C_2H_6FSi	$-SiF(CH_3)_2$	0.12	0.17	
C_2H_6NS	$-SN(CH_3)_2$	0.12	0.09	0.94
C_2H_6OP	$-PO(CH_3)_2$	0.43	0.50	2.81
$C_2H_6O_3P$	$-PO(OCH_3)_2$	0.42	0.55	
$C_2H_6O_4P$	$-OPO(OCH_3)_2$		0.04	
C_2; *5 elements*				
$C_2F_6NO_4S_2$	$-N(SO_2CF_3)_2$	0.75	0.80	4.4
C_2HClF_3O	$-OCF_2CHClF$	0.35	0.28	
C_2HF_3NO	$-NHCOCF_3$	0.30	0.12	
$C_2HF_4O_2S$	$-SO_2CF_2CHF_2$		1.01	
C_2H_2(halogen)N_2O_4	$-CH_2C(NO_2)_2$(halogen)			1.50
$C_2H_2F_3OS$	$-CH_2SOCF_3$	0.25	0.24	1.62
$C_2H_2F_3O_2S$	$-CH_2SO_2CF_3$	0.29	0.31	1.75
C_2H_3ClNO	$-NHCOCH_2Cl$	0.17	−0.03	2.06
C_2H_6ClNP	$-P(Cl)N(CH_3)_2$	0.38	0.56	
C_2H_6NOS	$-S(O)N(CH_3)_2$			1.87
$C_2H_6NO_2S$	$-SO_2N(CH_3)_2$	0.51	0.65	2.62
	$-N(CH_3)SO_2CH_3$	0.29	0.24	2.1
$C_2H_6NO_4S$	$-N(SO_2CH_3)_2$	0.47	0.49	2.8
C_2; *6 elements*				
$C_2HClF_3O_2S$	$-SO_2CF_2CHClF$		0.98	
$C_2H_3F_3NO_2S$	$-N(CH_3)SO_2CF_3$	0.46	0.44	3.0
C_3; *2 elements*				
C_3F_3	$-C\equiv CCF_3$	0.41	0.51	1.94
C_3F_5	$-CF=CF(CF_3)$(*trans*)	0.15	0.17	
C_3F_7	$-CF(CF_3)_2$	0.37	0.53	3.00
C_3H_3	$-C\equiv CCH_3$	0.10	0.12	1.20
	$-CH_2C\equiv CH$	0.07		0.81

Table A.1 (*Contd.*)

Substituent		σ_{meta}	σ_{para}	σ^*
C_3; *2 elements* (*Contd.*)				
C_3H_5	$-cyclo-C_3H_5$	−0.07	−0.21	
	$-C(CH_3)=CH_2$		0.16	
	$-CH=CHCH_3$	0.05	0.02	0.36
	$-CH_2CH=CH_2$	−0.11	−0.14	0.00
C_3H_6	$-3,4-(CH_2)_3-$	(−0.26)		
C_3H_7	$-CH(CH_3)CH_3$	−0.07	−0.15	−0.19
	$-CH_2CH_2CH_3$	−0.05	−0.15	−0.12
C_3; *3 elements*				
C_3F_6N	$-NC(CF_3)_2$	0.29	0.23	2.2
C_3HN_2	$-CH(CN)_2$	0.53	0.52	3.4
C_3HF_6	$-CH(CF_3)_2$			1.32
$C_3H_2F_3$	$-CH=CH(CF_3)$(*cis*)	0.16	0.17	
$C_3H_2F_3$	$-CH=CH(CF_3)$(*trans*)	0.20	0.20	
C_3H_2N	$-CH=CH(CN)$	0.24	0.17	
$C_3H_3Cl_2$	$-CH_2CH=CCl_2$	−0.06		
$C_3H_3N_2$	1-pyrazolyl			2.2
C_3H_3O	$-CH=CHCHO$	0.24	0.13	
$C_3H_3O_2$	$-CH=CHCOOH$	0.14	0.90	1.00
$C_3H_4Cl_3$	$-CH_2CH_2CCl_3$			0.25
C_3H_4N	$-CH_2CH_2CN$			0.87
C_3H_4O	$-CH_2CH_2COO^-$			0.02
C_3H_5O	$-CH_2COCH_3$			0.62
	$-COC_2H_5$	0.38	0.48	
$C_3H_5O_2$	$-CH_2OCOCH_3$	0.04	0.05	1.00
	$-CH_2CH_2COOH$	−0.03	−0.07	0.35
	$-COOC_2H_5$	0.37	0.45‡§	2.26
	$-CH_2COOCH_3$	0.13		1.06
C_3H_5S	$-SCH_2CH=CH_2$			1.45
$C_3H_5S_3$	$-SC(S)SC_2H_5$			3.00
C_3H_6O	$3,4-O(CH_2)_3O-$	(0.00)		
C_3H_7O	$-CH_2OC_2H_5$			0.58
	$-CH_2CH(OH)CH_3$	−0.12		−0.06
	$-OCH(CH_3)CH_3$	0.05	−0.45	1.62
	$-OCH_2CH_2CH_3$	0.10	−0.25	1.68
C_3H_7S	$-SCH(CH_3)CH_3$	0.23	0.07	1.56
	$-SCH_2CH_2CH_3$	0.22		1.49
C_3H_8N	$-CH_2N(CH_3)_2$	0.00	0.01	
C_3H_9Ge	$-Ge(CH_3)_3$	0.0	0.0	
C_3H_9N	$-CH_2NH(CH_3)_2^+$	0.40	0.43	
	$-NH_2CH_2CH_2CH_3^+$	0.71		3.74
	$-N(CH_3)_3^+$	0.99	0.96	4.55
C_3H_9P	$-P(CH_3)_3^+$	0.50	0.80	2.5
C_3H_9Si	$-Si(CH_3)_3$	0.11	0.0	−0.81
C_3H_9Sn	$-Sn(CH_3)_3$	0.0	0.0	
C_3; *4 elements*				
C_3HF_6O	$-C(OH)(CF_3)_2$	0.35	0.30	1.75

Table A.1 (*Contd.*)

Substituent		σ_{meta}	σ_{para}	σ^*
C_3; *4 elements* (*Contd.*)				
$C_3HF_6S_2$	$-CH(SCF_3)_2$	0.44	0.44	2.75
$C_3H_5N_2O_4$	$-C(C_2H_5)(NO_2)_2$	0.56	0.64	
$C_3H_5OS_2$	$-SC(S)OC_2H_5$			2.75
C_3H_6NO	$-CH_2NHCOCH_3$	−0.04	−0.05	0.43
	$-N(CH_3)COCH_3$	0.31	0.26	2.25
	$-CH_2CH_2CONH_2$	−0.06		0.19
	$-CON(CH_3)_2$			1.94
	$-ON=C(CH_3)_2$	0.29		1.81
	$-NHCOC_2H_5$	0.23		1.56
$C_3H_6NO_2$	$-OCON(CH_3)_2$			2.87
	$-C(CH_3)_2NO_2$	0.18	0.20	
	$-NHCOOC_2H_5$	0.07	−0.15	1.99
$C_3H_6NS_2$	$-SC(S)N(CH_3)_2$			2.31
	$-N(CH_3)C(S)SCH_3$			2.9
$C_3H_7N_2O$	$-NHCONHC_2H_5$	0.04	−0.26	
$C_3H_7N_2S$	$-NHCSNHC_2H_5$	0.30	0.07	
	$-NHCSN(CH_3)_2$			1.75
$C_3H_7O_2S$	$-SO_2CH(CH_3)CH_3$			3.68
	$-SO_2CH_2CH_2CH_3$	0.69		3.68
C_3H_9OSi	$-Si(CH_3)_2(OCH_3)$	0.04	−0.02	
	$-OSi(CH_3)_3$	0.13	−0.27	
$C_3H_9O_2Si$	$-Si(CH_3)(OCH_3)_2$	0.04	0.10	
$C_3H_9O_3Si$	$-Si(OCH_3)_3$	0.09	0.13	
C_3; *5 elements*				
$C_3HF_6O_2S$	$-SO_2CF_2CHFCF_3$		1.03	
$C_3H_2F_3O_2S$	$-CH=CHSO_2CF_3$	0.31	0.55	1.94
$C_3H_3F_3NO$	$-N(CH_3)COCF_3$	0.41	0.39	
C_3H_6NOS	$-SCON(CH_3)_2$			2.06
	$-OC(S)N(CH_3)_2$			3.12
C_4; *2 elements*				
C_4F_7	$-cyclo-C_4F_7$	0.49	0.53	2.81
C_4F_9	$-(CF_2)_3CF_3$	0.52	0.52	2.44
	$-C(CF_3)_3$	0.35	0.52‡	
C_4H_4	$-3,4-(CH)_4-$	(0.04†)		
C_4H_7	$-CH=CHC_2H_5$	−0.04		0.31
	$-CH_2CH=CHCH_3$	−0.04		0.00
	$-cyclo-C_4H_7$	−0.13	−0.15	
	$-CH=C(CH_3)_2$	−0.06		0.19
C_4H_8	$-3,4-(CH_2)_4-$	(−0.48)		
C_4H_9	$-(CH_2)_3CH_3$	−0.07	−0.16	−0.25
	$-CH(CH_3)C_2H_5$	−0.08	−0.19	−0.19
	$-C(CH_3)_3$	−0.09	−0.15	−0.30
	$-CH_2CH(CH_3)CH_3$	−0.07	−0.12	−0.19
C_4N_3	$-C(CN)_3$	0.98	0.99	6.1

Table A.1 (*Contd.*)

Substituent		σ_{meta}	σ_{para}	σ^*
C_4; *3 elements*				
$C_4F_9S_3$	$-C(SCF_3)_3$	0.51	0.53	3.06
C_4HN_2	$-CH=C(CN)_2$	0.55	0.70	2.56
$C_4H_2F_7$	$-CH_2C_3F_7$	0.08		0.87
$C_4H_3N_2$	$-C(CN)_2CH_3$	0.60	0.57	3.94
	2-pyrimidinyl			1.67
C_4H_3O	2-furyl	0.06	0.02†	0.25
	3-furyl			0.62
C_4H_3S	2-thienyl	0.09	0.05	1.31
	3-thienyl	0.03	−0.02	0.62
$C_4H_4Cl_3$	$-CH_2CH=CHCCl_3$			0.19
C_4H_4N	1-pyrrolyl	0.47	0.37	
C_4H_5O	$-CH=CHCOCH_3$	0.21	−0.01†	
C_4H_7O	$-COCH(CH_3)_2$	0.38	0.47	
$C_4H_7O_2$	$-CH_2COOC_2H_5$			0.82
	$-CH_2CH_2COOCH_3$			0.26
C_4H_7S	$-SCH_2CH_2CH=CH_2$			1.39
C_4H_9O	$-O(CH_2)_3CH_3$	−0.05	−0.32	1.68
	$-OCH(CH_3)C_2H_5$	0.25		1.62
	$-CH_2C(OH)(CH_3)_2$	−0.16		−0.25
$C_4H_9O_3$	$-C(OCH_3)_3$	−0.03	−0.04	
C_4H_9S	$-S(CH_2)_3CH_3$	0.21		1.44
	$-SCH(CH_3)C_2H_5$			1.32
$C_4H_{10}N$	$-NH(CH_2)_3CH_3$	−0.34	−0.51	−1.08
	$-N(C_2H_5)_2$	−0.15	−0.53	
$C_4H_{10}P$	$-P(C_2H_5)_2$		0.03	
$C_4H_{11}N$	$-NH_2(CH_2)_3CH_3^+$	0.71		3.74
	$-NH_2C(CH_3)_3^+$	0.71		
	$-CH_2CH_2NH(CH_3)_2^+$	0.24	0.14	
	$-NH_2CH_2CH(CH_3)_2^+$			3.60
	$-CH_2N(CH_3)_3^+$	0.40	0.44	1.90
$C_4H_{11}Si$	$-CH_2Si(CH_3)_3$	−0.17	−0.27	−0.31
C_4; *4 elements*				
$C_4F_6IO_4$	$-I(OCOCF_3)_2$	1.28	1.34	7.6
C_4HF_6O	$-cyclo-C(OH)(CF_2)_3$	0.49	0.53	2.81
$C_4H_6IO_4$	$-I(OCOCH_3)_2$	0.85	0.88	5.1
$C_4H_6NO_2$	$-N(COCH_3)_2$	0.35	0.33	2.31
C_4H_8NO	$-CH_2CH_2NHCOCH_3$			0.23
	$-(CH_2)_3CONH_2$	−0.08		0.12
	$-NHCOCH(CH_3)_2$	0.11	−0.10	
	$-CH_2CH_2CONHCH_3$			0.25
	-morpholino			0.69
$C_4H_8NO_2$	$-NHCH_2COOC_2H_5$	−0.10	−0.68	
$C_4H_{10}O_2P$	$-P(OC_2H_5)_2$		0.33	
$C_4H_{10}O_3P$	$-P(O)(OC_2H_5)_2$	0.49	0.57	3.02
$C_4H_{11}NO_2$	$-NH(CH_2CH_2OH)_2^+$			4.43
$C_4H_{11}OSi$	$-CH_2OSi(CH_3)_3$	−0.04	−0.05	

Table A.1 (*Contd.*)

Substituent		σ_{meta}	σ_{para}	σ^*
C_4; *4 elements* (*Contd.*)				
$C_4H_{12}N_2P$	$-P[N(CH_3)_2]_2$	−0.03	−0.06	
C_5; *2 elements*				
C_5H_7	−cyclo−1−pentenyl	−0.06	−0.05	
C_5H_9	$-cyclo-C_5H_9$	−0.15	−0.02	
C_5H_{11}	$-(CH_2)_2CH(CH_3)_2$		−0.23	
	$-CH_2C(CH_3)_3$	−0.13	−0.12	−0.12
	$-(CH_2)_4CH_3$	−0.08	−0.15	−0.23
	$-C(CH_3)_2C_2H_5$		−0.19	
C_5N_3	$-C(CN)=C(CN)_2$	0.77	0.98	4.2
C_5; *3 elements*				
$C_5H_3O_2$	2−furoyl	−0.05		0.25
C_5H_4N	2−pyridyl	0.33	0.17	
C_5H_5S	2−thenyl	−0.04		0.31
$C_5H_6N_3$	$2,4-(CH_3)_2-1,3,5-$triazinyl	0.25	0.39	
C_5H_7O	2−oxocyclopentyl			0.28
	3−oxocyclopentyl			0.19
$C_5H_7O_2$	$-CH=CHCOOC_2H_5$	0.19	0.03	
C_5H_9O	$-O-cyclo-C_5H_9$	0.25		1.62
	$-COC(CH_3)_3$	0.27	0.33	
C_5H_9S	$-S(CH_2)_3CH=CH_2$			1.36
$C_5H_{10}N$	−piperidino		−0.12	
$C_5H_{11}O$	$-O(CH_2)_4CH_3$	0.1	−0.34	0.95
$C_5H_{11}S$	$-S(CH_2)_4CH_3$			1.35
$C_5H_{12}N$	$-(CH_2)_3N(CH_3)_2$		−0.13	
$C_5H_{13}N$	$-CH_2CH_2N(CH_3)_3^+$	0.16	−0.01	
$C_5H_{13}Si$	$-CH_2CH_2Si(CH_3)_3$	−0.16		−0.25
C_5; *4 elements*				
$C_5H_{15}OSi_2$	$-Si(CH_3)_2OSi(CH_3)_3$	0.0	0.01	−0.81
C_6; *2 elements*				
C_6Cl_5	$-C_6Cl_5$	0.24	0.24	1.56
C_6F_5	$-C_6F_5$	−0.12	−0.03	
C_6H_5	$-C_6H_5$	0.05	−0.01	0.75
C_6H_9	−cyclo−1−hexenyl	−0.10	−0.08	
C_6H_{11}	−cyclohexyl	−0.14	−0.22	−0.18
C_6H_{13}	$-(CH_2)_5CH_3$	−0.16		−0.25
C_6; *3 elements*				
C_6H_4Br	$-(C_6H_4-2-Br)$			1.07
	$-(C_6H_4-3-Br)$		0.09	
	$-(C_6H_4-4-Br)$		0.08	0.86
C_6H_4Cl	$-(C_6H_4-2-Cl)$			1.05
	$-(C_6H_4-3-Cl)$			0.98
	$-(C_6H_4-4-Cl)$		0.08	0.87

Table A.1 (*Contd.*)

Substituent		σ_{meta}	σ_{para}	σ^*
C_6; *3 elements*				
C_6H_4F	$-(C_6H_4-3-F)$			0.95
	$-(C_6H_4-4-F)$			0.81
C_6H_4I	$-(C_6H_4-2-I)$			1.08
	$-(C_6H_4-3-I)$			0.90
	$-(C_6H_4-4-I)$			0.87
$C_6H_4N_3$	1, 2, 3 − benzotriazol − 2 − yl	0.49	0.51	
$C_6H_5N_2$	$-N=N-C_6H_5$	0.29	0.31†‡	1.87
C_6H_5O	$-(C_6H_4-4-OH)$		−0.24	
	$-OC_6H_5$	0.25	−0.32	2.43
$C_6H_5O_2$	$-O(C_6H_4-2-OH)$			2.60
C_6H_5S	$-(C_6H_4-2-SH)$			0.72
	$-SC_6H_5$	0.17	0.13	1.87
C_6H_5Se	$-SeC_6H_5$		0.13	2.30
C_6H_6N	$-(C_6H_4-4-NH_2)$		−0.30	
	$-NHC_6H_5$	−0.12	−0.45	
C_6H_7N	$-NH_2C_6H_5^+$			4.37
C_6H_8N	1 − (2, 5 − dimethyl)pyrrolyl	0.49	0.39	
$C_6H_{11}O$	$-O-$cyclo$-C_6H_{11}$	0.29		1.81
$C_6H_{11}S$	$-S-$cyclo$-C_6H_{11}$	0.31		1.93
$C_6H_{11}Se$	$-Se-$cyclo$-C_6H_{11}$	0.41		2.37
$C_6H_{13}S$	$-S(CH_2)_5CH_3$			1.33
$C_6H_{14}N$	$-(CH_2)_4N(CH_3)_2$	−0.08	−0.16	
$C_6H_{15}Ge$	$-Ge(C_2H_5)_3$	0.0	0.0	
$C_6H_{15}N$	$-(CH_2)_3N(CH_3)_3^+$	0.06	−0.01	
$C_6H_{15}Si$	$-Si(C_2H_5)_3$		0.0	
$C_6H_{15}Sn$	$-Sn(C_2H_5)_3$	0.0	0.0	
C_6; *4 elements*				
$C_6H_2N_3O_6$	$-[C_6H_2-2,4,6-(NO_2)_3]$	0.27	0.31	1.62
$C_6H_3Cl_2O$	$-O(C_6H_3-2,4-Cl_2)$			3.17
$C_6H_3N_2O_4$	$-[C_6H_3-2,4-(NO_2)_2]$			1.88
	$-[C_6H_3-3,5-(NO_2)_2]$			1.37
C_6H_4BrO	$-O(C_6H_4-2-Br)$			2.45
	$-O(C_6H_4-3-Br)$			2.48
	$-O(C_6H_4-4-Br)$			2.44
C_6H_4BrS	$-S(C_6H_4-3-Br)$			1.84
	$-S(C_6H_4-4-Br)$			1.83
C_6H_4BrSe	$-Se(C_6H_4-2-Br)$			1.60
	$-Se(C_6H_4-4-Br)$			1.42
C_6H_4ClO	$-O(C_6H_4-2-Cl)$			2.69
	$-O(C_6H_4-3-Cl)$			2.57
	$-O(C_6H_4-4-Cl)$			2.62
C_6H_4ClS	$-S(C_6H_4-2-Cl)$			2.12
	$-S(C_6H_4-3-Cl)$			2.02
	$-S(C_6H_4-4-Cl)$			1.97
C_6H_4ClSe	$-Se(C_6H_4-2-Cl)$			1.62
	$-Se(C_6H_4-3-Cl)$			1.51
	$-Se(C_6H_4-4-Cl)$			1.45

Table A.1 (*Contd.*)

Substituent		σ_{meta}	σ_{para}	σ^*
C_6; *4 elements* (*Contd.*)				
C_6H_4FO	$-O(C_6H_4-2-F)$			2.50
	$-O(C_6H_4-3-F)$			2.51
	$-O(C_6H_4-4-F)$			2.44
C_6H_4FS	$-S(C_6H_4-3-F)$			1.87
	$-S(C_6H_4-4-F)$			1.77
C_6H_4IO	$-O(C_6H_4-2-I)$			2.38
	$-O(C_6H_4-3-I)$			2.44
	$-O(C_6H_4-4-I)$			2.39
$C_6H_4NO_2$	$-(C_6H_4-2-NO_2)$			1.14
	$-(C_6H_4-3-NO_2)$		0.18	1.21
	$-(C_6H_4-4-NO_2)$		0.23	1.26
$C_6H_4NO_3$	$-O(C_6H_4-2-NO_2)$			2.78
	$-O(C_6H_4-3-NO_2)$			2.76
	$-O(C_6H_4-4-NO_2)$			2.91
C_6H_5ClP	$-P(Cl)C_6H_5$		0.44	
C_6H_5FP	$-P(H)(C_6H_4-3-F)$	0.09		
C_6H_5OS	$-S(O)C_6H_5$	0.51	0.46	3.24
$C_6H_5O_2S$	$-SO_2C_6H_5$	0.62	0.70	3.25
$C_6H_5O_3S$	$-S(O)_2OC_6H_5$		0.51	
	$-OSO_2C_6H_5$	0.36	0.33	3.62
C_6H_7NS	$-S(C_6H_4-4-NH_3^+)$			2.32
$C_6H_{14}O_3P$	$-P(O)(OC_3H_7)_2$	0.38	0.50	
$C_6H_{15}O_3Si$	$-Si(OC_2H_5)_3$	0.02	0.08	
$C_6H_{18}N_3Si$	$-Si[N(CH_3)_2]_3$	−0.04	−0.04	
C_6; *5 elements*				
C_6H_4BrOS	$-S(O)(C_6H_4-3-Br)$			3.12
	$-S(O)(C_6H_4-4-Br)$			3.14
$C_6H_4BrO_2S$	$-SO_2(C_6H_4-3-Br)$			3.35
	$-SO_2(C_6H_4-4-Br)$			3.35
C_6H_4ClFP	$-P(Cl)(C_6H_4-3-F)$	0.42		
C_6H_4ClOS	$-S(O)(C_6H_4-3-Cl)$			3.14
	$-S(O)(C_6H_4-4-Cl)$			3.14
$C_6H_4ClO_2S$	$-SO_2(C_6H_4-3-Cl)$			3.45
	$-SO_2(C_6H_4-4-Cl)$			3.49
C_6H_4FOS	$-S(O)(C_6H_4-3-F)$			3.11
	$-S(O)(C_6H_4-4-F)$			3.15
$C_6H_4FO_2S$	$-SO_2(C_6H_4-3-F)$			3.48
	$-SO_2(C_6H_4-4-F)$			3.40
$C_6H_4NO_2S$	$-S(C_6H_4-2-NO_2)$			2.47
	$-S(C_6H_4-3-NO_2)$			2.03
	$-S(C_6H_4-4-NO_2)$			2.33
$C_6H_4NO_2Se$	$-Se(C_6H_4-2-NO_2)$			1.84
	$-Se(C_6H_4-3-NO_2)$			1.65
	$-Se(C_6H_4-4-NO_2)$			1.83
$C_6H_4NO_3S$	$-S(O)(C_6H_4-3-NO_2)$			3.20
	$-S(O)(C_6H_4-4-NO_2)$			3.24
$C_6H_4NO_4S$	$-SO_2(C_6H_4-4-NO_2)$			3.63

Table A.1 (*Contd.*)

Substituent		σ_{meta}	σ_{para}	σ^*
C_6; *5 elements* (*Contd.*)				
$C_6H_6NO_2S$	$-SO_2NHC_6H_5$	0.56	0.65	
	$-NHSO_2C_6H_5$	0.16	0.01	1.99
C_6; *6 elements*				
C_6H_4ClFOP	$-PO(Cl)(C_6H_4-3-F)$	0.65		
C_6H_4ClFPS	$-PS(Cl)(C_6H_4-3-F)$	0.56		
C_7; *2 elements*				
C_7H_7	$-CH_2C_6H_5$	−0.08	−0.09	0.27
	$-(C_6H_4-2-CH_3)$			0.62
	$-(C_6H_4-4-CH_3)$		−0.05	0.59
C_7H_{13}	$-CH_2-cyclo-C_6H_{11}$	−0.17		−0.31
C_7H_{15}	$-(CH_2)_6CH_3$	−0.19		−0.37
C_7; *3 elements*				
C_7H_5O	$-COC_6H_5$	0.36	0.46	2.2
$C_7H_5O_2$	$-OCOC_6H_5$	0.21	0.13	2.57
	$-COOC_6H_5$	0.37	0.44	
$C_7H_5O_3$	$-O(C_6H_4-2-COOH)$			3.20
	$-OCO(C_6H_4-4-OH)$			2.4
C_7H_6N	$-N=CHC_6H_5$	−0.08	−0.55	
	$-CH=NC_6H_5$	0.35	0.42	
C_7H_7O	$-OCH_2C_6H_5$		−0.41	
	$-CH(OH)C_6H_5$	0.00	−0.03	0.50
	$-CH_2(C_6H_4-4-OH)$		0.93	
	$-(C_6H_4-4-OCH_3)$		−0.09	0.60
	$-O(C_6H_4-2-CH_3)$			2.29
	$-O(C_6H_4-3-CH_3)$			2.33
	$-O(C_6H_4-4-CH_3)$			2.30
	$-CH_2OC_6H_5$	0.04	0.05	0.87
$C_7H_7O_2$	$-O(C_6H_4-2-OCH_3)$			2.29
	$-O(C_6H_4-3-OCH_3)$			2.42
	$-O(C_6H_4-4-OCH_3)$			2.32
C_7H_7S	$-SCH_2C_6H_5$	0.23		1.56
	$-S(C_6H_4-2-CH_3)$			1.90
	$-S(C_6H_4-3-CH_3)$			1.89
	$-S(C_6H_4-4-CH_3)$			1.80
$C_7H_7S_2$	$-S(C_6H_4-2-SCH_3)$			1.62
	$-S(C_6H_4-3-SCH_3)$			1.93
	$-S(C_6H_4-4-SCH_3)$			1.69
C_7H_7Se	$-Se(C_6H_4-2-CH_3)$			1.33
	$-Se(C_6H_4-3-CH_3)$			1.30
	$-Se(C_6H_4-4-CH_3)$			1.23
C_7H_9N	$-(C_6H_4-3-CH_2NH_3^+)$			1.39
	$-(C_6H_4-4-CH_2NH_3^+)$			1.32
$C_7H_{13}O$	$-OCH_2-cyclo-C_6H_{11}$	0.18		1.31

Table A.1 (*Contd.*)

Substituent		σ_{meta}	σ_{para}	σ^*
C_7; *4 elements*				
$C_7H_4ClO_2$	$-OCO(C_6H_4-4-Cl)$			2.63
$C_7H_4F_3S$	$-S(C_6H_4-3-CF_3)$			2.02
C_7H_4NO	$-O(C_6H_4-2-CN)$			2.67
	$-O(C_6H_4-3-CN)$			2.59
	$-O(C_6H_4-4-CN)$			2.73
	2-benzoxazolyl	0.30	0.33	
$C_7H_4NO_4$	$-OCO(C_6H_4-4-NO_2)$			2.73
C_7H_4NS	2-benzthiazolyl	0.27	0.29	
	$-S(C_6H_4-4-CN)$			2.29
C_7H_6NO	$-CONHC_6H_5$	0.23	0.41	1.56
	$-NHCOC_6H_5$	0.22	0.08	1.68
$C_7H_6NO_2$	$-CH_2(C_6H_4-4-CN)$			0.45
C_7H_7FP	$-P(CH_3)(C_6H_4-3-F)$	0.20		
$C_7H_7N_2O$	$-N=N(C_6H_3-2-OH-5-CH_3)$	0.27	0.31	
C_7H_7OS	$-S(C_6H_4-2-OCH_3)$			1.59
	$-S(C_6H_4-3-OCH_3)$			1.89
	$-S(C_6H_4-4-OCH_3)$			1.66
	$-S(O)(C_6H_4-3-CH_3)$			2.99
	$-S(O)(C_6H_4-4-CH_3)$			3.02
C_7H_7OSe	$-Se(C_6H_4-2-OCH_3)$			1.17
	$-Se(C_6H_4-3-OCH_3)$			1.38
	$-Se(C_6H_4-4-OCH_3)$			1.18
$C_7H_7O_2S$	$-S(O)(C_6H_4-4-OCH_3)$			3.00
	$-SO_2(C_6H_4-3-CH_3)$			3.27
	$-SO_2(C_6H_4-4-CH_3)$			3.32
$C_7H_7O_3S$	$-SO_2(C_6H_4-3-OCH_3)$			3.24
	$-SO_2(C_6H_4-4-OCH_3)$			3.23
	$-O(C_6H_4-4-SO_2CH_3)$			2.85
C_7H_7SSe	$-Se(C_6H_4-2-SCH_3)$			1.27
	$-Se(C_6H_4-4-SCH_3)$			1.23
C_7H_8OP	$-P(OCH_3)C_6H_5$		0.32	
$C_7H_{14}NO$	$-CH_2CH_2CONHC(CH_3)_3$			0.25
$C_7H_{21}O_2Si_3$	$-Si(CH_3)[OSi(CH_3)_3]_2$	−0.07	0.02	
C_7; *5 elements*				
C_7H_7FOP	$-P(OCH_3)(C_6H_4-3-F)$	0.33		
	$-PO(CH_3)(C_6H_4-3-F)$	0.40		
C_8; *2 elements*				
C_8H_5	$-C\equiv CC_6H_5$	0.14	0.16	1.35
C_8H_7	$-CH=CHC_6H_5$	0.03	−0.07	0.41
C_8H_9	$-(C_6H_4-4-C_2H_5)$			0.59
	$-CH(CH_3)C_6H_5$	−0.03		0.37
	$-CH_2CH_2C_6H_5$	−0.07	−0.12	−0.06

Table A.1 (*Contd.*)

Substituent		σ_{meta}	σ_{para}	σ^*
C_8; *3 elements*				
C_8H_6N	$-CH_2(C_6H_4-4-CN)$			0.41
	3-indolyl	−0.12		−0.06
$C_8H_7O_2$	$-O(C_6H_4-4-COCH_3)$			2.91
$C_8H_7O_3$	$-OCO(C_6H_4-4-OCH_3)$			2.50
C_8H_9O	$-CH(OCH_3)C_6H_5$		−0.01	
C_8H_9S	$-CH_2SCH_2C_6H_5$	−0.03		0.38
	$-SCH_2CH_2C_6H_5$	0.21		1.44
$C_8H_{11}Si$	$-Si(CH_3)_2C_6H_5$	0.04	0.07	−0.87
$C_8H_{17}O$	$-O(CH_2)_5CH(CH_3)_2$		−0.27	
C_8; *4 elements*				
C_8H_6NO	$-CH=NCOC_6H_5$	0.39	0.51	
C_8H_8NO	$-N(COCH_3)C_6H_5$	0.19		1.37
	$-CH_2CONHC_6H_5$	−0.11		0.00
	$-N=CH(C_6H_4-4-OCH_3)$	−0.07	−0.54	
$C_8H_8NO_2$	$-NHCO(C_6H_4-4-OCH_3)$	0.09	−0.06	
	$-OCO(C_6H_4-4-NHCH_3)$			~2.0
C_8H_9OSe	$-Se(C_6H_4-2-OC_2H_5)$			1.13
	$-Se(C_6H_4-4-OC_2H_5)$			1.18
$C_8H_9O_3S$	$-CH_2OSO_2(C_6H_4-4-CH_3)$			1.44
$C_8H_{18}OP$	$-P(O)(C_4H_9)_2$	0.35	0.49	
$C_8H_{18}O_3P$	$-P(O)(OC_4H_9)_2$			1.77
C_8; *5 elements*				
$C_8H_8NO_2S_2$	$-S-CH=NSO_2(C_6H_4-4-CH_3)$	0.65	0.70	
C_9; *2 elements*				
C_9H_{11}	$-[C_6H_4-4-CH(CH_3)_2]$			0.56
	$-(CH_2)_3C_6H_5$	−0.12		−0.06
C_9; *3 elements*				
C_9H_7O	$-CH=CHCOC_6H_5$	0.18	0.05	
$C_9H_{13}N$	$-[C_6H_4-3-N(CH_3)_3]^+$			1.65
	$-[C_6H_4-4-N(CH_3)_3]^+$			1.51
C_9; *4 elements*				
$C_9H_6NO_3$	$-CH=CHCO(C_6H_4-4-NO_2)$	0.15	0.05	
$C_9H_{11}N_2O_6$	uridin−5−yl			~0.8
$C_9H_{27}O_3Si_4$	$-Si[OSi(CH_3)_3]_3$	−0.09	−0.01	
C_{10}; *2 elements*				
$C_{10}H_5$	$-C\equiv CC\equiv CC_6H_5$			2.81
$C_{10}H_7$	α-naphthyl	0.06		0.75
	β-naphthyl	0.06		0.75

Table A.1 (*Contd.*)

Substituent		σ_{meta}	σ_{para}	σ^*
C_{10}; *2 elements* (*Contd.*)				
$C_{10}H_{13}$	$-[C_6H_4-4-C(CH_3)_3]$			0.52
$C_{10}H_{15}$	adamantyl	−0.12	−0.24	
$C_{10}H_{19}$	$-(CH_2)_4-cyclo-C_6H_{11}$	−0.19		−0.37
C_{10}; *3 elements*				
$C_{10}H_9Fe$	Ferrocenyl	−0.15	−0.18	
$C_{10}H_{13}S$	$-S[C_6H_4-4-C(CH_3)_3]$			1.50
$C_{10}H_{17}O$	decahydro − 2 − naphthyloxy			1.68
C_{10}; *4 elements*				
$C_{10}H_{13}OS$	$-S(O)[C_6H_4-4-C(CH_3)_3]$			2.97
$C_{10}H_{13}O_2S$	$-SO_2[C_6H_4-4-C(CH_3)_3]$			3.23
C_{11}; *2 elements*				
$C_{11}H_9$	$-CH_2-(1-C_{10}H_7)$	−0.01		0.44
C_{12}; *3 elements*				
$C_{12}H_{10}N$	$-N(C_6H_5)_2$	−0.07	−0.28	
$C_{12}H_{10}P$	$-P(C_6H_5)_2$	0.11	0.19	1.06
$C_{12}H_{27}P$	$-P(C_4H_9)_3{}^+$			3.61
C_{12}; *4 elements*				
$C_{12}H_{10}NO$	$-N(COCH_3)(1-C_{10}H_7)$	0.25		1.62
	$-N(COCH_3)(2-C_{10}H_7)$	0.26		1.68
$C_{12}H_{10}OP$	$-PO(C_6H_5)_2$	0.40	0.54	1.71
$C_{12}H_{10}PS$	$-PS(C_6H_5)_2$	0.29	0.47	
C_{13}; *2 elements*				
$C_{13}H_{11}$	$-CH(C_6H_5)_2$	−0.03	−0.04	0.41
C_{13}; *3 elements*				
$C_{13}H_9N_2$	1 − phenyl − 2 − benzimidazolyl	0.07	0.12	
$C_{13}H_{11}S_2$	$-CH(SC_6H_5)_2$			1.56
$C_{13}H_{13}Si$	$-Si(CH_3)(C_6H_5)_2$	0.10	0.13	
C_{13}; *4 elements*				
$C_{13}H_{10}NO$	$-CON(C_6H_5)_2$		0.35	
C_{14}; *3 elements*				
$C_{14}H_{11}O_2$	$-COOCH(C_6H_5)_2$	0.36	0.55	
C_{15}; *2 elements*				
$C_{15}H_{15}$	$-CH(C_6H_4-2-CH_3)_2$			0.40

Table A.1 (*Contd.*)

Substituent		σ_{meta}	σ_{para}	σ^*
C_{18}; *3 elements*				
$C_{18}H_{15}Ge$	$-Ge(C_6H_5)_3$	0.05	0.08	
$C_{18}H_{15}P$	$-P(C_6H_5)_3{}^+$			4.46
$C_{18}H_{15}Si$	$-Si(C_6H_5)_3$	−0.03	0.10	
C_{19}; *2 elements*				
$C_{19}H_{15}$	$-C(C_6H_5)_3$	−0.01	0.02	~0.4
C_{19}; *3 elements*				
$C_{19}H_{15}S$	$-SC(C_6H_5)_3$	0.05		0.69
C_n				
OR	−OR	0.1	−0.35	1.69
SR	−SR	0.18	0.05	1.56
HNR	−NHR	−0.29	−0.56	
H_2NR	$-NH_2R^+$			3.75
CO_2R	−COOR	0.35	0.44	1.94
O_2PR_2	$-P(OR)_2$	0.12	0.15	
O_2SR	$-SO_2R$	0.64	0.73	3.56
O_3PR_2	$-PO(OR)_2$	0.38	0.50	
O_3SR	$-SO_2OR$	0.71	0.90	3.12
CHO_2R_2	$-CH(OR)_2$	−0.04	−0.05	1.12
CHS_2R_2	$-CH(SR)_2$			0.94
CH_2OR	$-CH_2OR$	0.02	0.02	
$C_2H_2O_2R$	$-CH_2COOR$			1.12
$C_2H_2O_3R$	$-OCH_2COOR$		−0.18	
$C_3H_2O_2R$	$-CH=CH-COOR$	0.19	0.03	1.12
$CHNO_2R$	−NHCOOR	0.07	−0.15	
CH_2O_2SR	$-CH_2SO_2R$	0.15	0.17	1.12
CH_3NO_2R	$-NHCH_2COOR$	−0.10	−0.68	

† If a phenol, see Table A.4 (where the value for a phenol is not available, try the value for an aniline).
‡ If an aniline, see Table A.4 (where the value for an aniline is not available, try the value for a phenol).
§ If a pyridine, see Table A.4 (where the value for a pyridine is not available, use σ_{para} of Table A.1).

A.2 Some Hammett and Taft Equations for pK_a values of Organic Acids at 25° C. (Table A.2)

The grouping, RC_6H_4-, in a molecular formula can be taken as subsuming substitution in the benzene ring by more than one −R, unless otherwise implied by the absence of Σ in the appropriate equation.

Some equations relate to acids involving aryl groups. In the cases where aryl equals phenyl, the sigma constant(s) for use in the equations is(are) for the substituent(s) attached to the benzene ring. For polycyclic aromatic compounds, the substituent constants for the annelated rings and for the substituents they carry must also be taken into account.

Acids	pK_a =
Acetoacetic esters: $CH_3COCH(R)COOC_2H_5$	$12.59 - 3.44\,\sigma^{*a}$ (Barlin and Perrin, 1966)
Acetylacetones (enol form): $CH_3COC(R) = C(OH)CH_3$	$9.25 - 1.78\,\sigma^*$ (Barlin and Perrin, 1966)
Acrylic acids: $H_2C = C(R)COOH$	$4.27 - 4.25\,\sigma_{meta}$
$(CH_3)_2C{=}C(R)COOH$	$4.62 - 2.70\sigma_{meta}$ (Charton, 1965)
$RHC = CHCOOH$, *trans*	$4.39 - 2.23\,\sigma_{para}$ (Charton and Meislich, 1958)
$RHC{=}CHCOOH$, *cis*	$4.04 - 2.34\,\sigma_{para}$
$R(CH_3)C{=}CHCOOH$, *cis*	$4.62 - 3.29\,\sigma_{para}$ (Charton, 1965a)
$R(CH_3)C{=}CHCOOH$, *trans*	$4.61 - 2.98\,\sigma_{para}$
$R(COOH)C{=}CHCOOH$, *trans*	$5.02 - 1.92\,\sigma_{para}$ (Charton and Meislich, 1958)
$R(Cl)C{=}CHCOOH$, *cis*	$3.56 - 1.91\,\sigma_{para}$
$RHC{=}C(CH_3)COOH$, *cis*	$3.78 - 3.48\,\sigma_{para}$
$RHC{=}C(CH_3)COOH$, *trans*	$4.61 - 2.61\,\sigma_{para}$ (Charton, 1965a)
Alcohols: RCH_2OH	$15.9 - 1.42\,\sigma^*$ (Ballinger and Long, 1960)
	$15.74 - 1.32\,^{*b}$ (Takahashi *et al.*, 1971)
$R'R''CH(OH)$	$15.9 - 1.42\,\Sigma\sigma^*$ (Ballinger and Long, 1960)
Aldehydes, hydrated: $RCH(OH)_2$	$14.4 - 1.42\,\sigma^*$ (Barlin and Perrin, 1966)
Aliphatic carboxylic acids: $RCOOH$	$4.66 - 1.62\,\sigma^*$ (Barlin and Perrin, 1966)
RCH_2COOH	$4.76 - 0.67\,\sigma^{*c}$ (Barlin and Perrin, 1972)
$R^1R^{11}CHCOOH$	$4.80 - 0.66\,\Sigma\sigma^*$
$R^1R^{11}R^{111}COOH$	$5.10 - 0.81\,\Sigma\sigma^*$ (This work)
Amides: R^1CONHR^{11}	$22.0 - 3.1\,\Sigma\sigma^{*d}$ (Barlin and Perrin, 1966)
Arenearsonic acids: $ArAsO(OH)_2$	$3.54 - 1.05\,\Sigma\sigma$ (pK_{a1})
	$8.49 - 0.87\,\Sigma\sigma$ (pK_{a2}) (Jaffé, 1953)
Areneboronic acids: $ArB(OH)_2$	$9.70 - 2.15\,\Sigma\sigma^{e}$ (Jaffé, 1953)

Table A.2 (*Contd.*)

Acids		pK_a =
Arenephosphonic acids: $ArPO(OH)_2$		$1.84 - 0.76\ \Sigma\sigma\ (pK_{a1})$ $6.97 - 0.95\ \Sigma\sigma\ (pK_{a2})$ (Jaffé, 1953)
—, 2-chloro (or bromo-):		$2.94 - 0.99\ \Sigma\sigma^{f}\ (pK_{a1})$ $8.25 - 1.19\ \Sigma\sigma^{f}\ (pK_{a2})$
In water:		$6.90 - 0.91\ \Sigma\sigma\ (pK_{a2})$ (McDaniel and Brown, 1958)
Areneseleninic acids: ArSeOOH		$4.74 - 0.91\ \Sigma\sigma$ (Gould and McCullough, 1951)
Arenesulphonic acids: $ArSO_3H$		$-6.7 - 0.7\ \Sigma\sigma$ (Maarsen *et al.*, 1974)
Aromatic hydroxamic acids: ArCONHOH		$8.78 - 1.02\ \Sigma\sigma$ (This work)
Arylacetic acids: $ArCH_2COOH$, ring substituted		$4.30–0.49\ \Sigma\sigma$ (McDaniel and Brown, 1958)
3-Arylacrylic acids: Ar(H)C = CHCOOH, ring substituted		$4.45–0.47\ \Sigma\sigma$ (McDaniel and Brown, 1958)
α-Arylaldoximes: ArCH = N(OH), ring substituted		$10.70–0.86\ \Sigma\sigma$ (Jaffé, 1953)
Arylcarboxylic acids: ArCOOH		$4.20–1.00\ \Sigma\sigma$ $5.24–1.09\ \Sigma\sigma^{g}$ $5.08–1.42\ \Sigma\sigma^{h}$ $7.21–1.96\ \Sigma\sigma^{i}$ $4.21–0.96\ \Sigma\sigma^{j}$
—, 2-nitro-:		$2.21–0.90\ \Sigma\sigma$
—, 2-methyl-:		$3.88–1.43\ \Sigma\sigma$
—, 2-hydroxy-:		$4.00–1.10\ \Sigma\sigma$
—, 2-chloro-:		$3.69–0.85\ \Sigma\sigma$
—, 2,6-dimethyl-:		$3.97–1.12\ \Sigma\sigma^{k}$
—, 4-aryl-: substitution in the ring carrying −COOH		$5.64–0.48\ \Sigma\sigma^{l}$ (McDaniel and Brown, 1958)
2-Arylcyclopropyl carboxylic acids, aryl ring substituted	*cis:*	$6.33–0.44\ \Sigma\sigma^{m}$
	trans:	$5.78–0.47\ \Sigma\sigma^{m}$ (Fuchs *et al.*, 1962)
(Arylthio) acetic acids: $ArSCH_2COOH$, ring substituted		$3.57–0.30\ \Sigma\sigma$ (Pasto *et al.*, 1965)
Benzamidates, *N*-trimethylammonio-: $RC_6H_4CONH\overset{+}{N}(CH_3)_3$		$4.26–1.51\ \Sigma\sigma$ (Beck and Liler, 1978)
Benzenesulphonamides:		$10.00–1.06\ \Sigma\sigma^{n}$ (Willi, 1956)
—, $R.C_6H_4SO_2NH_2$		$10.02–0.88\ \sigma$
—, $R^1R^{11}.C_6H_3SO_2NH_2$		$10.05–0.93\ \Sigma\sigma$
—, $C_6H_5SO_2NH(C_6H_4)R$		$8.99–2.04.\sigma^{o}$
—, *m*-$NO_2.C_6H_4SO_2NHC_6H_4R$		$7.88–1.98\ \sigma^{o}$
—, $RC_6H_4SO_2NHCH_3$		$11.43–1.44\ \sigma^{o}$
—, $C_6H_5CH_2SO_2NHC_6H_4R$		$9.35–1.77\ \sigma$

Table A.2 (*Contd.*)

Acids	pK_a =
Benzenesulphonamides: (*Contd.*)	
—, $RC_6H_4SO_2NHCH_2C_6H_5$	11.21–1.95 σ (Dauphin and Kergomard, 1961)
Benzenesulphonanilides: $R.C_6H_4SO_2NHC_6H_5$	8.31–1.16 σ
—, $C_6H_5SO_2NHC_6H_4.R$	8.31–1.74 σ (Willi, 1956)
Benzoic acids:	4.20–1.00 $\Sigma\sigma$ (Hammett, 1940)
—, 2-chloro-	3.69–0.86 $\Sigma\sigma$
—, 2-hydroxy-	4.00–1.10 $\Sigma\sigma$
—, 2-methyl-	3.90–1.22 $\Sigma\sigma$
—, 2-nitro-	2.21–0.91 $\Sigma\sigma$ (Shorter and Stubbs, 1949)
Benzimidazoles, 2-substituted:	8.54–7.92 σ_{meta} (Charton, 1965d)
1,4-Benzoquinones, 3,6-(R, R^1)-2,5-dihyroxy-:	2.38–7.04 $\Sigma\sigma_{para}$ (pK_{a1}) 4.63–6.72 $\Sigma\sigma_{para}$ (pK_{a2}) (Charton, 1965a)
Butanoic acids, 4-(R)-4-oxo-: $RCOCH_2CH_2COOH$	4.61–1.31 σ_{meta} (Charton, 1965)
2-Butenoic acids, 2-substituted, *cis*: $CH_3CH=C(R)COOH$	4.19–4.44 σ_{meta}
—, 2-substituted, *trans*:	4.70–3.92 σ_{meta} (Charton, 1965)
3-Butenoic acids, 4-aryl-2-oxo- $ArCH=CHCOCOOH$	1.97–0.05 $\Sigma\sigma$ 2.60–0.17 $\Sigma\sigma^p$ (McDaniel and Brown, 1958)
Cinnanamides, N-trimethylammonio-, ring substituted: $RC_6H_4CH=CH-CONHN(CH_3)_3^+$	4.57–0.70 $\Sigma\sigma$ (Beck and Liler, 1978)
Dinitromethanes: $RC(NO_2)_2H$	5.35–2.23 σ^{*q} (This work)
Ethanols, 2,2,2-trifluoro-1-monosubstituted phenyl: $F_3CCH(OH)C_6H_4R$	11.90–1.01 σ (Stewart and van der Linden, 1960)
2-Furancarboxylic acids:	3.13–1.40 $\Sigma\sigma$ (Kwok *et al.*, 1964)
Methanohydroxamic acids: RCONHOH	9.48–0.98 σ^* (Swidler *et al.*, 1959)
1,4-Naphthoquinones, 3-substituted-2-hydroxy-:	5.16–1.40 σ^* (Barlin and Perrin, 1966) 4.37–4.55 σ_{para} (Charton, 1965a)
Oxazoles, 2-amino-4-(*p*-substituted) phenyl-:	3.43–1.29 $\sigma_{para}{}^r$ (Strauss *et al.*, 1973)

Table A.2 (*Contd.*)

Acids		pK_a =
Oximes: $R^1R^{11}C=NOH$		12.35–1.18 $\Sigma\sigma^{*}$ [s]
$R^1(CH_3CO)C=NOH$		9.00–0.94 σ^* (Barlin and Perrin, 1972)
Oximinocarboxylic acids, *anti* substituted: $HON=C(R)COOH$		3.03–2.37 σ_{meta} (Charton, 1965)
Phenols:		9.92–2.23 $\Sigma\sigma$ (Biggs and Robinson, 1961)
Phenoxyacetic acids: $RC_6H_4OCH_2COOH$		3.18–0.23 $\Sigma\sigma$ (Pasto *et al.*, 1965)
		3.17–0.30 $\Sigma\sigma$ (Pettit *et al.*, 1968)
Phenylacetamidates, N-trimethylammonio-: $RC_6H_4CH_2CONH\overset{+}{N}(CH_3)_3$		4.68–0.64 $\Sigma\sigma$ (Beck and Liler, 1978)
Phenylacetic acids: $RC_6H_4CH_2COOH$		4.30–0.49 $\Sigma\sigma$ (Jaffé, 1953)
3-Phenylcyclobutanecarboxylic acids: aryl ring substituted,	*cis*:	5.98–0.26 $\Sigma\sigma$ [t] (Caputo and Fuchs, 1968)
2-Phenylcyclopropanecarboxylic acids: aryl ring substituted,	*cis*:	6.38–0.44 $\Sigma\sigma$ [t]
	trans:	5.78–0.47 $\Sigma\sigma$ [t] (Fuchs *et al.*, 1962)
3-Phenylpropanoic acids, ring substituted:		4.55–0.21 $\Sigma\sigma$ (Jaffé, 1953)
		4.71–0.24 $\Sigma\sigma$ (Pasto *et al.*, 1965)
3-Phenylpropenoic acids, *cis*, 2-(R): $C_6H_5CH=C(R)COOH$		3.70–4.42 σ_{meta} (Charton, 1965)
		4.14–0.66 σ_{para} (Charton, 1965a)
—, *trans*, 2-(R):		4.64–3.76 σ_{meta} (Charton, 1965)
—, *trans*, 3-(R):		4.45–3.87 σ_{para} (Charton, 1965a)
Phenylpropiolic acids: $RC_6H_4C\equiv CCOOH$		3.24–0.81 $\Sigma\sigma$ [u] (Newman and Merrill, 1955)
(Phenylseleno)acetic acids: $RC_6H_4SeCH_2COOH$		3.75–0.35 $\Sigma\sigma$ (Pettit *et al.*, 1968)
Phenylselenonic acids: $RC_6H_4SeO_2OH$		4.78–1.03 $\Sigma\sigma$ (Gould and McCullough, 1951)
Phenylsulphinylacetic acids: $RC_6H_4SO.CH_2COOH$		2.73–0.17 $\Sigma\sigma$ (Pasto *et al.*, 1965)
Phenylsulphonylacetic acids: $RC_6H_4SO_2.CH_2COOH$		2.51–0.25 $\Sigma\sigma$ (Pasto *et al.*, 1965)

Table A.2 (*Contd.*)

Acids	pK_a =
(Phenylthio)acetic acids: $RC_6H_4SCH_2COOH$	3.38–0.32 $\Sigma\sigma$ (Pettit *et al.*, 1968)
Propanoic acids, 3-(R)-3-oxo: $RCOCH_2COOH$	3.51–3.34 σ_{meta} (Charton, 1965)
Propenoic acids, 3-aryl-, *trans*: ArCH = CHCOOH	4.45–0.47 $\Sigma\sigma$ (Dippy and Lewis, 1937)
Propiolic acids: RC ≡ CCOOH	2.39–1.89 σ_{para} (Charton, 1961)
2-Pyridones:	11.65–4.28 $\Sigma\sigma$ (Barlin and Perrin, 1966)
—, 3-substituted:	12.35–0.79 σ^* (Spinner and White, 1966)
4-Pyridones:	11.12–4.28 $\Sigma\sigma$ (Barlin and Perrin, 1966)
4-Pyrones, 3-hydroxy-6-substituted:	8.01–1.63 σ_{para} (Choux and Benoit, 1967)
Pyrroles:	17.00–4.28 $\Sigma\sigma$ (Barlin and Perrin, 1966)
Pyrrole-2-carboxylic acids:	4.50–1.65 $\Sigma\sigma$ (Fringuelli *et al.*, 1969)
—, 5-substitued:	2.82–1.40 σ_{para} (Jaffé, 1953)
Selenophen-2-carboxylic acids:	3.60–1.23 $\Sigma\sigma$ (Fringuelli *et al.*, 1972)
Tellurophen-2-carboxylic acids:	3.97–1.20 $\Sigma\sigma$ (Fringuelli *et al.*, 1972)
Tetronic acids, α-substituted:	3.76–2.23 σ_{ortho} [v] (Barlin and Perrin, 1972)
	3.39–3.75 σ_{para} (Charton, 1965*a*)
Thiazoles, 2-amino-4-(*p*-substituted)phenyl-:	8.00–7.3 σ_{meta} (Freiberg and Kruger, 1967)
—, 2-amino-4-(*p*-substituted)phenyl-:	4.00–1.24 σ_{para} [w] (Stauss *et al.*, 1973)
—, 2-amino-5-(*p*-substituted)phenyl-:	4.38–1.18 σ_{para} [w]
Thiobenzoic acids: $RC_6H_4.COSH$	2.61–1.0 $\Sigma\sigma$ (Barlin and Perrin, 1966)
Thioformic acids: RCOSH	3.52–1.63 σ^* (Barlin and Perrin, 1966)
Thiols: RSH	10.22–3.50 σ^*
RCH_2SH	10.54–1.47 σ^* (Kreevoy *et al.*, 1960, 1964)
Thiophen-2-carboxylic acids:	3.53–1.20 $\Sigma\sigma$ (Fringuelli *et al.*, 1972)
—, 4-substituted:	3.53–0.97 σ
—, 5-substituted:	3.53–1.10 σ (Otsuji *et al.*, 1959)

Table A.2 (*Contd.*)

Acids	pK_a =
Thiophenols:	6.52–2.2 $\Sigma\sigma$ (Barlin and Perrin, 1966)
1,2,4-Triazoles, 3-nitro-5-substitued:	5.95–8.92 σ_{meta} (Bagal and Pevsner, 1970)
α,α,α-Trifluoroacetophenones, hydrated: $RC_6H_4C(OH)_2CF_3$	10.00–1.11 σ (Stewart and van der Linden, 1960)
Tropolones, 5-substituted:	6.42–3.10 σ_{para} (Oka *et al.*, 1962)

[a] For ionization of a proton from –COCH(R)CO–
[b] Add 0.50 if carbon-2 is sp^2 hybridized
[c] The first member of this series is anomalous. Moreover, the pK_a is not significantly changed if one of the α-hydrogens is replaced by an alkyl substituent. If both α-hydrogens are replaced, the pK_a is raised by 0.3 of a unit.
[d] Constants are for R^1CO- and -R^{11}.
[e] In aqueous 25% ethanol.
[f] In aqueous 50% ethanol.
[g] In aqueous 50% methanol.
[h] In aqueous 50% ethanol.
[i] In 100% ethanol.
[j] In water at 35°.
[k] In aqueous 20% dioxan.
[l] In aqueous 50% butyl cellosolve.
[m] In aqueous 50% ethanol.
[n] At 20°, $I = 0.1$.
[o] In aqueous 23.4% ethanol.
[p] In aqueous 50% ethanol.
[q] Where $-C(NO_2)_2H$ is joined to an aliphatic carbon in R.
[r] In aqueous 50% ethanol.
[s] R^1,R^{11} are not acyl groups.
[t] In aqueous 50% ethanol.
[u] In aqueous 35% dioxan.
[v] The σ_{ortho} constants are for phenols.
[w] In aqueous 50% ethanol.

A.3 Some Hammett and Taft Equations for pK_a Values of Protonated Bases at 25° C (Table A.3)

The grouping, RC_6H_4-, in a molecular formula can be taken as subsuming substitution in the benzene ring by more than one –R, unless otherwise implied by the absence of Σ in the appropriate equation.

Protonated bases	pK_a =
Acetophenones, ring substituted, protonated on 0:	$-6.0-2.6\ \Sigma\sigma$ (Clark and Perrin, 1964).
Amidinium ions: $RC(=\overset{+}{N}H_2)NH_2$	$11.46-11.98\ \sigma_{meta}$ (Charton, 1965b)
—, N^1,N^1-dibutyl-C–aryl:	$11.14-1.41\ \sigma^a$ (Jaffé, 1953)
—, N-phenyl-C-substituted:	$8.90-12.08\ \sigma_{meta}$ (Charton, 1965b)
Aminium ions, primary: RNH_3^+	$10.15-3.14\ \sigma^*$ (Hall, 1957)
—, primary: $RCH_2NH_3^+$	$10.4-1.83\ \sigma^*$ (This work)
—, secondary: $R'R''NH_2^+$	$10.59-3.23\ \Sigma\sigma^*$ (Hall, 1957)
—, secondary: $(R'CH_2)(R''CH_2)NH_2^+$	$10.4-2.05\ \Sigma\sigma^*$ (This work)
—, tertiary: $R'R''R'''NH^+$	$9.61-3.30\ \Sigma\sigma^*$ (Hall, 1957)
—, tertiary: $(R'CH_2)(R''CH_2)(R'''CH_2)NH^+$	$10.7-1.83\ \Sigma\sigma^*$ (This work)
Anilinium ions: $RC_6H_4NH_3^+$	$4.58-2.88\ \Sigma\sigma$ (Biggs and Robinson, 1961)
—, N,N-dimethyl-: $RC_6H_4N(CH_3)_2H^+$	$5.06-3.46\ \Sigma\sigma$ (Taft and Lewis, 1959)
1–Azoniabicyclo[2.2.2]octane ions, 3–substituted:	$11.05-0.90\ \sigma^*$ (Grob *et al.*, 1957)
Azobenzenes, 4-substituted, protonated:	$-2.2-3.0\ \sigma$
—, 4,4′-substituted, protonated:	$0.01-4.3\ \Sigma\sigma$ (This work)
Benzimidazolinium ions, 2-substituted:	$5.58-10.9\ \sigma_{meta}$ (Charton, 1965d)
Benzoic acids, protonated: $RC_6H_4COOH_2^+$	$-7.26-1.2\ \Sigma\sigma$ (Clark and Perrin, 1964)
Benzylaminium ions, ring substituted:	$9.39-1.05\ \Sigma\sigma$ (Blackwell *et al.*, 1964)
2,2′-Bipyridylium monocations, 4,4′-di(R):	$4.2-2.7\sigma_{para}+0.4\sigma_{meta}$ (Tsai, 1967)
1,2-Dihydro-2-oxopyridinium ions:	$1.1-4.3\ \Sigma\sigma^b$ (This work)
1,4-Dihydro-4-oxopyridinium ions:	$3.2-2.86\ \Sigma\sigma^c$ (This work)
N,N-Diethylaniline oxides, protonated:	$4.26-0.91\ \Sigma\sigma$ (Perrin, 1980)
N,N-Dimethylaniline oxides, protonated:	$4.50-0.91\ \Sigma\sigma$ (Perrin, 1980)
Guanidinium ions: $RNHC(=\overset{+}{N}H_2)NH_2$	$14.0-3.60\ \sigma^*$ (This work)

Table A.3 (*Contd.*)

Protonated bases	pK_a =
Hydrazinium ions: $RNHNH_3^+$	7.80–2.80 σ^* (This work)
Imidazolinium ions, 2-substituted:	7.08–10.9 σ_{meta}
—, 1-methyl-4-substituted:	7.12–10.6 σ_{meta} (Charton, 1965d)
Isoquinolinium ions, 1-, 3- and 4-substituted:	5.32–5.90 $\Sigma\sigma$ (Clark and Perrin, 1964)
Isoquinolinium ions, 1-substituted:	4.22–8.69 σ_{meta} (Charton, 1964)
—, 4-substituted:	5.41–5.36 σ_{meta}
—, 5-substituted:	5.40–2.49 σ_{meta}
—, 6-substituted:	5.45–2.57 σ_{para}
—, 7-substituted:	5.71–2.89 σ_{meta}
—, 8-substituted:	5.52–2.75 σ_{meta} (Charton, 1965c)
Naphth[2,3]imidazolinium ions,2-substituted:	5.31–11.0 σ_{meta}
—, 5,6,7,8-tetrahydro-2-substituted:	6.01–8.87 σ_{meta} (Charton, 1965d)
α-Naphthylaminium ions, 2-, 3- and 4-substituted:	3.85–2.81 $\Sigma\sigma$ (Clark and Perrin, 1964)
β-Naphthylaminium ions, 3- and 4-substituted:	4.29–2.81 $\Sigma\sigma^d$ (Clark and Perrin, 1964)
3-Nitroguanidinium ions: $RNHC(=\overset{+}{N}H_2)NHNO_2$	10.89–8.28 σ^e (Charton, 1965b)
N-Phenylamidinium ions: $RC(=\overset{+}{N}H_2)NHC_6H_5$	8.90–12.08 σ_{meta} (Charton, 1965b)
Phenylhydrazinium ions: $RC_6H_4NHNH_3^+$	8.73–1.21 $\Sigma\sigma$ (Stroh and Westphal, 1963)
—, α-methyl-: $RC_6H_4N(CH_3)NH_3^+$	9.02–1.86 $\Sigma\sigma$ (Stroh and Westphal, 1964)
Phosphinium ions, secondary: $R'R''PH_2^+$	3.59–2.61 $\Sigma\sigma^*$
—, tertiary: $R'R''R'''PH^+$	7.85–2.67 $\Sigma\sigma^*$ (Henderson and Streuli, 1960)
Pyridazinium ions, 3-dimethylamino-6-alkyl-:	5.14–6.14 σ^f
—, 3-chloro-6-substituted:	−0.21–6.7 σ_{para} (Cookson and Cheeseman, 1972)
3-Pyridazones, 6-alkyl-, protonated:	−1.22–2.92 σ_{para} (Cookson and Cheeseman, 1972)
Pyridine 1-oxides, protonated:	0.94–2.09 $\Sigma\sigma$ (Jaffe and Doak, 1955)
4-Pyridinethiones, 2-substituted-5-methoxy-, protonated:	1.68–3.16 σ_{meta}
—, 2-substituted-5-methoxy-1-methyl-, protonated:	1.54–3.99 σ_{meta} (Besso *et al.*, 1977)

Table A.3 (*Contd.*)

Protonated bases	pK_a =
Pyridinium ions:	5.25–5.90 $\Sigma\sigma$ (Clark and Perrin, 1964)
	5.39–5.70 $\Sigma\sigma$ (Charton, 1964)
—, 2-substituted:	5.03–11.8 σ_{meta} (Charton, 1964)
—, 2-substituted-5-methoxy-4-methylthio-:	5.54–7.33 σ_{ortho} (Besso *et al.*, 1977)
—, 5-ethyl-2-styryl-, substituted, *trans*:	5.57–0.48 $\Sigma\sigma$
—, 2-styryl-, substitued, *cis*:	5.18–0.29 $\Sigma\sigma$
trans:	5.18–0.57 $\Sigma\sigma$
—, 4-styryl-, substituted, *trans*:	4.97–0.47 $\Sigma\sigma$ (Doty *et al.*, 1969)
Pyrimidinium ions:	1.23–5.90 $\Sigma\sigma$ (This work)
—, 2-substituted:	0.64–3.66 σ_{para}
—, 4-substituted:	1.38–5.85 σ_{para}
—, 2-methylamino-4-substituted:	3.89–5.72 σ_{para}
—, 4-amino-6-substituted:	5.47–9.38 σ_{meta}
—, 2,4-diamino-5-substituted:	7.66–6.38 σ_{meta}
—, 2,4-diamino-6-substituted:	6.79–7.98 σ_{meta} (Roth and Strelitz, 1969)
—, 2-methyl-4-amino-5-substituted:	6.52–5.39 σ
—, 2-methyl-4-dimethylamino-5-substituted:	7.51–5.53 σ
—, 2-methyl-4,5-disubstituted:	3.12–5.01 $\Sigma\sigma$ (Mizukami and Hirai, 1966)
Quinolinium ions, 2-, 3- and 4-substituted:	4.84–5.90 $\Sigma\sigma$ (Clark and Perrin, 1964)
—, 2-substituted:	5.12–9.04 σ_{meta} (Charton, 1964)
—, 5-substituted:	5.03–3.44 σ_{meta}
—, 6-substituted:	5.16–3.34 σ_{meta}
—, 7-substituted:	4.80–2.96 σ_{para}
—, 8-substituted:	4.64–3.11 σ_{meta} (Charton, 1965c)
Semicarbazones, $R'R''C=NNHCONH_2$ protonated:	1.3–0.35 $\Sigma\sigma^*$ (This work)

[a] In aqueous 50% ethanol.
[b] Protonation takes place on the oxygen atom, so that positions 3,4 and 5 are taken respectively as *ortho*, *meta* and *para* to the basic centre.
[c] Protonation takes place on the oxygen atom so that positions, 2, 3 and 5 are taken respectively as *meta*, *ortho* and *ortho* to the basic centre.
[d] Fails for 1– and 3–NO_2.
[e] .I = 1.0.
[f] In aqueous 50% methylcellosolve.

A.4 Special σ Constants for Para Substituents

Substituent	Phenols	Anilines	Pyridines
C_0			
$-O^-$	−0.66		
$-ClO_3$	1.03		
$-OH$	0.03	−0.31	0.35
$-SH$			0.65
$-NH_2$	−0.29		
$-NH_3^+$		0.56	
$-NO_2$	1.24	1.26	1.26
$-SO_3^-$	0.39	0.46	0.31
$-SO_2NH_2$		0.8	
C_1			
$-CF_3$		0.74	
$-CN$	0.88	1.00	0.56
$-CHO$	1.03	0.99	0.99
$-OCH_3$	−0.11		−0.21
$-SCH_3$	0.21	0.08	−0.12
$-SO_2CH_3$	0.92	1.14	
C_2			
$-COCH_3$	0.84	0.81	0.28
$-COOCH_3$		0.75	
$-N(CH_3)_2$	−0.12		
C_3			
$-COOC_2H_5$		0.72	
C_4			
$-3,4\text{-}(CH)_4-$	0.11		
C_6			
$-N=N-C_6H_5$	0.77	0.57	
C_7			
$-COC_6H_5$		0.83	

A.5 Apparent σ_{ortho} Constants for use in Hammett Equations

Substituent	Carboxylic Acids	Phenols	Anilines	Pyridines
C_0				
$-Br$	1.35	0.70	0.71	0.58
$-Cl$	1.28	0.68	0.67	0.79
$-F$	0.93	0.54	0.47	
$-I$	1.34	0.63	0.70	
$-O^-$		~ −1.1		
$-OH$	1.22	0.04	−0.09	0.76
$-SH$	~0.5			
$-NH_2$		0.03	0.00	−0.27
$-NH_3^+$	2.15		1.23	
$-NO_2$	1.99	1.40	1.72	
$-SO_2^-$		0.21	0.75	
$-PO(OH)O^-$	~0.4			
C_1				
$-CH_3$	0.29	−0.13	0.10	−0.13
$-CN$	1.06	1.32		
$-COO^-$	−0.91		−0.13	−0.05
$-CHO$		0.75		
$-COOH$	0.95			0.51
$-CH_2OH$		0.04		
$-OCH_3$	0.12	0.00	0.00	0.34
$-SCH_3$	0.52	~0.3		0.28
$-CH_2NH_3^+$		0.41		
$-CONH_2$	0.45	0.72		
$-SOCH_3$		1.04		
C_2				
$-C_2H_5$	0.41	−0.09	0.05	−0.13
$-OCH_2COO^-$	−0.27			
$-COCH_3$	0.07			
$-OCOCH_3$	−0.37			
$-COOCH_3$	0.63			0.51
$-OC_2H_5$	−0.01	−0.08	0.02	
$-N(CH_3)_2$		−0.36		
$-CH_2CH_2NH_3^+$		0.28		0.25
$-CH_2NH_2CH_3^+$				0.39
C_3				
$-cyclo-C_3H_5$	0.07			
$-CH(CH_3)CH_3$	0.56	−0.23	0.03	−0.11
$-CH_2CH_2CH_3$		−0.25		−0.14

Table A.5 (*Contd.*)

Substituent	Carboxylic Acids	Phenols	Anilines	Pyridines
C_3 (*Contd.*)				
$-CH=N-C_2H_5$		~ −0.85		
$-OCH(CH_3)CH_3$	−0.04			
$-OCH_2CH_2CH_3$	−0.04			
$-CH_2CH_2CH_2NH_3^+$		0.14		
$-CH_2NH(CH_3)_2^+$		0.54		
$-N(CH_3)_3^+$		1.07		
C_4				
$-2,3-(CH)_4-$	0.50	0.28	0.24	0.06
$-2,3-(CH_2)_4-$		−0.13		
$-CH_2CH_2CH_2CH_3$		~ −0.3		
$-C(CH_3)_3$	0.66	−0.52		−0.11
$-COCH_2CH_2CH_3$	~ −0.4			
$-CH_2CH_2NH(CH_3)_2^+$		0.48		
$-CH_2N(CH_3)_3^+$		0.48		
C_5				
$-CH_2C(CH_3)_3$		−0.39		
$-CH=N(CH_2)_3CH_3$		~ −0.95		
$-CH=NC(CH_3)_3$		~ −1.4		
$-CH_2CH_2N(CH_3)_3^+$		0.11		
C_6				
$-C_6H_5$	0.74	0.00		
$-(CH_2)_5CH_3$		~ −0.35		
$-OC_6H_5$	0.67			
$-SC_6H_5$			0.72	
$-NHC_6H_5$	~0.2			
$-(CH_2)_3N(CH_3)_3^+$		0.02		
$-N=N-(C_6H_4-2-OH)$		0.82		
C_7				
$-CH_2C_6H_5$				0.02
$-COC_6H_5$	0.65			
$-CH_2(C_6H_4-2-OH)$		0.95		
$-CH_2(C_6H_4-4-OH)$		0.73		
C_8				
$-CO(C_6H_4-4-CH_3)$	0.56			

A.6 Sigma Constants for Heteroatoms in Heterocyclic Rings

Five-membered rings

Heterogroup	Substituent on 6-membered ring†	σ_{ortho}	σ_{meta}
–O–	–OR	1.08‡	0.25‡
–S–	–SR	0.72‡	0.12‡
–NH–	–NHR	–0.24‡	–0.34‡
–Se–	–SeR	0.67‡	
–Te–	–TeR	0.23‡	

Six-membered rings

Heterogroup	σ_{ortho}	σ_{meta}	σ_{para}
$=NH^+$	3.21‡	2.18‡	4.06§
=N–	0.56§	0.73§	0.83§
$=\overset{+}{N}O$		1.48‡	1.35‡
$=\overset{+}{N}OH$		2.3 ¶	3.9§

† Using the empirical procedure described in Section 8.1.
‡ Derived from carboxylic acids.
§ From azapyridines.
¶ From anilines.

References

Adams, R. and Mahan, J. E. (1942), *J. Amer. Chem. Soc.*, **64**, 2588.
Albert, A. and Armarego, W. L. F. (1965), *Adv. Heterocyclic Chem.*, **4**, 1.
Albert, A. and Serjeant, E. P. (1971), *The Determination of Ionization Constants*, 2nd Edn, Chapman and Hall, London.
Arnett, E. M., Chawla, B., Bell, L., Taagepera, M., Hehre, W. J. and Taft, R. W. (1977), *J. Amer. Chem. Soc.*, **99**, 5729.
Bagal, L. I. and Pevzner, M. S. (1970), *Khim. Geterotsikl. Soedin.*, **6**, 558.
Ballinger, P. and Long, F. A. (1960), *J. Amer. Chem. Soc.*, **82**, 795.
Barlin, G. B. and Perrin, D. D. (1966), *Quarterly Rev.*, **20**, 75.
Barlin, G. B. and Perrin, D. D. (1972), in *Techniques of Chemistry–Vol. IV* (eds Bentley and Kirby), Wiley, New York.
Bates, R. G. (1973), *Determination of pH: Theory and Practice*, 2nd Edn, Wiley, New York.
Beck, W. H. and Liler, M. (1978), *J. Chem. Soc. Perkins Trans. II*, 1173.
Besso, H., Imapuku, K. and Matsumura, H. (1977), *Bull. Chem. Soc. Japan*, **50**, 3295.
Blackwell, L. F., Fischer, A., Miller, I. J., Topsom, R. D. and Vaughan, J. (1964), *J. Chem. Soc.*, 3588.
Biggs, A. I. and Robinson, R. A. (1961), *J. Chem. Soc. (London)*, 388.
Bolton, P. D., Hall, F. M. and Reece, I. H. (1966), *Spectrochim. Acta.*, **22**, 1149.
Brönsted, J. N. (1923), *Rec. trav. Chim.*, **47**, 718.
Bryson, A. (1960), *J. Amer. Chem. Soc.*, **82**, 4862.
Bryson, A. (1960a), *J. Amer. Chem. Soc.*, **82**, 4871.
Bunting, J. W. (1979), *Adv. Heterocyclic Chem.*, **25**, 1.
Bunting, J. W. and Perrin, D. D. (1966), *J. Chem. Soc., B*, 436.
Caputo, J. A. and Fuchs, R. (1968), *J. Org. Chem.*, **33**, 1959.
Charton, M. (1961), *J. Org. Chem.*, **26**, 735.
Charton, M. (1964), *J. Amer. Chem. Soc.*, **86**, 2033.
Charton, M. (1965), *J. Org. Chem.*, **30**, 557.
Charton, M. (1965a), *J. Org. Chem.*, **30**, 974.
Charton, M. (1965b), *J. Org. Chem.*, **30**, 969.
Charton, M. (1965c), *J. Org. Chem.*, **30**, 3341.
Charton, M. and Meislich, H. (1958), *J. Amer. Chem. Soc.*, **80**, 5940.
Chiang, Y. and Whipple, E. B. (1963), *J. Amer. Chem. Soc.*, **85**, 2763.
Choux, G. and Benoit, R. L. (1967), *J. Org. Chem.*, **32**, 3974.
Clark, J. and Perrin, D. D. (1964), *Quarterly Rev.*, **18**, 295.
Cookson, R. F. (1974), *Chem. Rev.*, **74**, 5.
Cookson, R. F. and Cheeseman, G. W. H. (1972), *J. Chem. Soc. Perkin Trans.*, **2**, 392.
Dauphin, G. and Kergomard, A. (1961), *Bull. Soc. Chim. France*, **3**, 486.

Davies, C. W. (1938), *J. Chem. Soc.*, 2093.
Dewar, M. J. S. and Grisdale, P. J. (1962), *J. Amer. Chem. Soc.*, **84**, 3539, 3548.
Doty, J. C., Williams, J. L. R. and Grisdale, P. J. (1969), *Can. J. Chem.*, **47**, 2355.
Edward, J. T., Chang, H. S., Yates, K. and Stewart, R. (1960), *Can. J. Chem.*, **38**, 1518.
Exner, O. (1972), in *Advances in Free Energy Relationships* (eds. Chapman and Shorter) Plenum Press, London.
Exner, O. (1978), in *Correlation Analysis in Chemistry* (eds. Chapman and Shorter) Plenum Press, London.
Fringuelli, F., Marino, G. and Savelli, G. (1969), *Tetrahedron*, **25**, 5815.
Fringuelli, F., Marino, G. and Tattichi, A. (1972), *J. Chem. Soc. Perkins Trans.*, **2**, 1738.
Fuchs, R., Kaplan, C. A., Bloomfield, J. J. and Hatch, L. F. (1962), *J. Org. Chem.*, **27**, 733.
Fuson, R. C. (1935), *Chem. Rev.*, **16**, 1.
Gore, P. H. and Phillips, J. N. (1949), *Nature*, **163**, 690.
Gould, E. S. and McCullough, J. D. (1951), *J. Amer. Chem. Soc.*, **73**, 1109.
Grob, C. A., Kaiser, A. and Renk, E. (1957), *Chem. Ind. (London)*, 598.
Hall, N. S. (1930), *J. Amer. Chem. Soc.*, **52**, 5115.
Hall, H. K. (1957), *J. Amer. Chem. Soc.*, **79**, 5441.
Hammett, L. P. (1935), *Chem. Rev.*, **17**, 125.
Hammett, L. P. (1940), *Physical Organic Chemistry*, McGraw-Hill, New York, Chap. 7.
Hansch, C., Leo, A., Unger, S. H., Kim, K. H., Nikaitani, D. and Lien, E. J. (1973), *J. Med. Chem.*, **16**, 1207.
Henderson, W. A. and Streuli, C. A. (1969), *J. Amer. Chem. Soc.*, **82**, 5791.
Jaffé, H. H. (1953), *Chem. Rev.*, **53**, 191.
Jaffé, H. H. and Doak, G. O. (1955), *J. Amer. Chem. Soc.*, **77**, 4441.
Jaffé, H. H. and Lloyd Jones, H. (1964), *Adv. Heterocyclic Chem.*, **3**, 209.
Jones, J. R. and Taylor, S. E. (1980), *J. Chem. Res. (S)*, 154.
King, E. J. (1965), *Acid-Base Equilibria*, Pergamon Press, Oxford.
Kirkwood, J. G. and Westheimer, F. H. (1938), *J. Phys. Chem.*, **6**, 506, 513.
Kirkwood, J. G. and Westheimer, F. H. (1939), *J. Amer. Chem. Soc.*, **61**, 555.
Kreevoy, M. M., Eichinger, B. E., Stary, F. A., Katz, E. A. and Sellsedt, J. H. (1964), *J. Org. Chem.*, **29**, 1641.
Kreevoy, M. M., Harper, E. T., Duvall, R. E., Wigus, H. S. and Dutsch, L. T. (1960), *J. Amer. Chem. Soc.*, **86**, 4899.
Kuthan, J., Danihel, I. and Skela, V. (1978), *Coll. Czech. Chem. Comm.*, **43**, 447.
Kwok, W. K., O'Ferrall, R. A. M. and Miller, S. I. (1964), *Tetrahedron*, **20**, 1913.
Longuet-Higgins, H. C. (1950), *J. Chem. Phys.*, **18**, 265, 275, 283.
Maarsen, P. K., Bregman, R. and Cerfontain, H. (1974), *Tetrahedron*, **30**, 1211.
McCullough, J. D. and Gould, E. S. (1949), *J. Amer. Chem. Soc.*, **71**, 674.
Mastryukova, T. A. and Kabachnik, M. I. (1969), *Russ. Chem. Rev.*, **38**, 795.

Mastryukova, T. A. and Kabachnik, M. I. (1971), *J. Org. Chem.*, **36**, 1201.
Mizukami, S. and Hirai, E. (1966), *J. Org. Chem.*, **31**, 1199.
Newman, M. S. and Merrill, S. H. (1955), *J. Amer. Chem. Soc.*, **77**, 5552.
Oka, Y., Umehara, M. and Nozoe, T. (1962), *Nippon Kagaku Zasshi*, **83**, 1197.
Otsuji, Y., Kimura, T., Suzimoto, Y. and Imoto, E., Omori, Y. and Okawara, T. (1959), *Nippon Kagaku Zasshi*, **80**, 1021.
Pasto, D. J. McMillan, D. and Murphy, T. (1965), *J. Org. Chem.*, **30**, 2688.
Perrin, D. D. (1964), *Austral. J. Chem.*, **17**, 484.
Perrin, D. D. (1965), *J. Chem. Soc.*, 5590.
Perrin, D. D. (1965a), *Adv. Heterocyclic Chem.*, **4**, 43.
Perrin, D. D. (1980), 'Prediction of pK_a values' in *Physical Chemical Properties of Drugs* (eds. Yalkowsky, S. H., Sinkula, A. A. and Valvani, C.) Marcel Dekker, New York, Chap. 1.
Pettit, L. D., Royston, A., Sherrington, C. and Whewell, R. J. (1968), *J. Chem. Soc., B*, 588.
Reynolds, W. F., Mezey, P. G., Hehre, W. J., Topsom, R. D. and Taft, R. W. (1977), *J. Amer. Chem. Soc.*, **99**, 5821.
Robinson, R. A. (1964), *J. Res. Nat. Bur. Stand.*, **A68**, 159.
Rochester, C. H. (1970), *Acidity Functions*, Academic Press, London.
Roth, B. and Strelitz, J. Z. (1969), *J. Org. Chem.*, **34**, 821.
Shorter, J. and Stubbs, F. J. (1949), *J. Chem. Soc.*, 1180.
Sommer, P. F., Pascual, C., Arya, V. P. and Simon, W. (1963), *Helv. Chim. Acta*, **46**, 1734.
Spinner, E. and White, J. C. B. (1966), *J. Chem. Soc. (London), B*, 991.
Stauss, U., Härter, H. P. and Schindler, O. (1973), *Chimia*, **27**, 99.
Stewart, R. and Granger, M. R. (1961), *Can. J. Chem.*, **39**, 3508.
Stewart, R. and Yates, K. (1958), *J. Amer. Chem. Soc.*, **80**, 6355.
Stewart, R. and Yates, K. (1960), *J. Amer. Chem. Soc.*, **82**, 4059.
Stewart, R. and van der Linden, R. (1960), *Can. J. Chem.*, **38**, 399.
Stroh, H. H. and Westphal, G. (1963), *Chem. Ber.*, **96**, 184.
Stroh, H. H. and Westphal, G. (1964), *Chem. Ber.*, **97**, 83.
Swidler, R., Flapinger, R. E. and Steinberg, G. M. (1959), *J. Amer. Chem. Soc.*, **81**, 3271.
Taft, R. W. and Lewis, I. C. (1959), *J. Amer. Chem. Soc.*, **81**, 5343.
Takahashi, S., Cohen, L. A., Miller, H. K. and Peake, E. G. (1971), *J. Org. Chem.*, **36**, 1205.
Tsai, C. S. (1967), *Can. J. Chem.*, **45**, 2862.
Willi, A. V. (1956), *Helv. Chim. Acta*, **39**, 46.
Yates, K. and Stewart, R. (1959), *Can J. Chem.*, **37**, 664.
Yoshioka, M., Hamamoto, K. and Kubota, T. (1962), *Bull. Chem. Soc., Japan*, **35**, 1723.

Index